SpringerBriefs in Computer Science

Series Editors
Stan Zdonik
Peng Ning
Shashi Shekhar
Jonathan Katz
Xindong Wu
Lakhmi C. Jain
David Padua
Xuemin Shen
Borko Furht
VS Subrahmanian
Martial Hebert
Katsushi Ikeuchi
Bruno Siciliano

For further volumes:
http://www.springer.com/series/10028

Haijun Zhang • Xiaoli Chu • Xiangming Wen

4G Femtocells

Resource Allocation and Interference Management

 Springer

Haijun Zhang
College of Information Science
    and Technology
Beijing University of Chemical
    Technology
Beijing, China, People's Republic

Xiangming Wen
Beijing University of Posts
    and Telecommunications
Beijing, China, People's Republic

Xiaoli Chu
Department of Electronic
    and Electrical Engineering
The University of Sheffield
Sheffield, UK

ISSN 2191-5768   ISSN 2191-5776 (electronic)
ISBN 978-1-4614-9079-1   ISBN 978-1-4614-9080-7 (eBook)
DOI 10.1007/978-1-4614-9080-7
Springer New York Heidelberg Dordrecht London

Library of Congress Control Number: 2013947965

Printed on acid-free paper

Springer is part of Springer Science+Business Media (www.springer.com)

Preface

Femtocells have been considered as a promising technology to provide better indoor coverage and spatial reuse gains in the last few years. Femtocells are low power, low cost and user deployed wireless access points that use local broadband connections as backhaul. Not only the users but also the operators benefit from femtocells. On the one hand, users enjoy high-quality links; on the other hand, operators decrease the operational expenditure (OPEX) and capital expenditure (CAPEX) due to the traffic offloading and user's self-deployment of femtocell base stations (FBSs). Orthogonal frequency division multiple access (OFDMA) based femtocells have been considered in major wireless communication standards, e.g., LTE/LTE-Advanced. Due to spectrum scarcity and implementation difficulty, spectrum-sharing, rather than spectrum splitting, between femtocells and macrocells is more preferable from the operator's perspective. However, co-channel deployed femtocells may lead to severe co-channel interference between femtocells in dense deployment, and cross-tier inference between macro-tier and femto-tier.

Due to the fading coefficients of different subchannels are likely to be independent for different users, which are known as multiuser diversity (MUD), maximum system spectral efficiency can be achieved by selecting the best user for each subchannel and adapting the associated transmit power. Therefore, resource allocation is one of the most important techniques for femtocells to maximize spectral efficiency and mitigate interference. Power control and subchannel allocation have been widely used to alleviate cross-tier and/or co-tier interference and satisfy diverse quality of service (QoS) for co-channel deployment of femtocells. However, there has not been any book specifically addressing femtocell network resource allocation with various objectives, constraints and optimizing variables taken into consideration.

In this book, we address the foregoing issues and provide an in-depth discussion on the latest resource allocation and interference management issues for femtocells. The discussion begins with introducing femtocells and their development in Chap. 1. After that, resource allocation in dense deployed femtocells is investigated in Chap. 2. Such techniques include user scheduling and power control to maximize capacity of femtocells. In Chap. 3, an interference-aware pricing-based resource allocation algorithm for co-channel femtocells is proposed to alleviate their interference to macrocells without degrading the femtocells' capacity. The subchannel and power allocation problem is modeled as a non-cooperative game. A suboptimal subchannel allocation algorithm and an optimal power allocation algorithm are proposed to implement the resource allocation game. In Chap. 4, resource allocation is investigated in both uplink and downlink for two-tier networks comprising spectrum-sharing femtocells and macrocells. A resource allocation scheme for co-channel femtocells is proposed, aiming to maximize the capacity for both delay-sensitive users and delay-tolerant users subject to delay-sensitive users' QoS constraints and the interference constraint imposed by the macrocell. The subchannel and power allocation problem is modeled as a mixed integer programming problem, then transformed into a convex optimization problem by relaxing subchannel sharing, and finally solved by the dual decomposition method. The complexity of the proposed algorithms is analyzed, and the effectiveness of the proposed algorithms is verified by simulations. In Chap. 5, we propose an energy-aware uplink power control scheme for two-tier femto-macro networks based on non-cooperative game. In Chap. 6, we propose a differentiated-pricing based power allocation algorithm for the uplink of spectrum-sharing femtocells, based on a non-cooperative game framework. Concluding remarks and future trends are provided in Chap. 7.

Beijing, China, People's Republic   Haijun Zhang
Sheffield, UK   Xiaoli Chu
Beijing, China, People's Republic   Xiangming Wen

Contents

Chapter 1
Introduction to 4G Femtocells

Abstract Femtocells have been proposed for improving the performance of indoor users to provide better indoor coverage and spatial reuse gains in the 4G networks. On the one hand, users enjoy high-quality links; on the other hand, operators decrease the operational expenditure (OPEX) and capital expenditure (CAPEX) due to the traffic offloading and user's self-deployment of femtocell base stations (FBSs). As femtocells can meet users' demand and indoor coverage requirement well, they have been widely used in many wireless communication standards, such as WiMAX, and LTE/LTE-Advanced. However, there are still some challenges in the mass deployed femtocell environment. Interference management is considered as one of the major challenges in femto-macro two-tier networks. In this chapter, we survey different state-of-the-art approaches of resource allocation and interference management in orthogonal frequency division multiple access (OFDMA) femtocell networks. Moreover, some open challenges in interference and resource management are discussed.

1.1 4G Femtocell Networks

Fourth-generation (4G) mobile networks are expected to provide high capacity and wide coverage. However, since the 4G wireless systems, such as WiMAX and LTE/LTE-Advanced, are usually deployed in high frequency band, the penetration loss will be high. Moreover, above 50% of voice services and 70% of data traffics occur indoors nowadays [1]. The most promising solution to this problem is shortening the distance between the transmitter and the receiver.

Insufficient indoor coverage of macrocells has led to increasing interest in femtocells, which have been proposed for improving the quality of service (QoS) of indoor users [2]. Femtocells usually comprise small size, low-power, low-cost, and short-range home base stations. They work in the licensed frequency bands, and are connected to broadband Internet backhaul. As femtocells can meet

H. Zhang et al., *4G Femtocells: Resource Allocation and Interference Management*,
SpringerBriefs in Computer Science, DOI 10.1007/978-1-4614-9080-7_1,
© The Author(s) 2013

customers' demands and indoor coverage requirements, femtocells combined with orthogonal frequency division multiple access (OFDMA) have been considered in many wireless communication standards, such as WiMAX and LTE/LTE-Advanced [3]. Two-tier OFDMA macrocell and femtocell networks are widely expected to improve coverage and capacity of indoor environments.

Dedicated-channel deployment of femtocells, where femtocells and macrocells are assigned with different (or orthogonal) frequency bands, may not be preferred by operators due to the scarcity of spectrum resources and complexities in implementation. While in co-channel deployment, where femtocells and macrocells share the same spectrum, cross-tier interference could be severe [2], especially when femtocell base stations (FBSs) are deployed close to a macrocell base station (MBS) [22]. Due to the fundamental role of macrocells in providing blanket cellular coverage, their capacities and coverage should not be affected by co-channel deployment of femtocells. As a result, resource allocation and interference management have become an important asset to enhance performance and have attracted much attention from the telecommunication industry.

1.2   Resource Allocation and Interference Management

In practice, there are still some technical challenges to be further addressed before extensive deployment of femtocells. A two-tier macrocell and femtocell network is usually implemented by sharing frequency rather than splitting frequency between tiers [4]. Hence, cross-tier interference (CTI) and inter-tier interference (ITI) are the key issues in two-tier macrocell and femtocell networks [1]. Maximization of the total data rate of femtocells with the consideration of cross-tier and inter-tier interference has become an interesting research area. Related works on femtocell networks in the literature are described in the following. Resource allocation algorithms aiming at the inter-cell interference management in femtocell networks are discussed and evaluated in [5]. A resource allocation scheme considering inter-femtocell fairness is proposed in [6]. In [7], a cross-tier interference mitigation algorithm based on power control is developed. The authors in [27] propose a distributed resource allocation algorithm based on Lagrangian dual method.

Power control has been widely used to mitigate inter-cell interference in co-channel deployment of femtocells. For alleviating uplink interference caused by co-channel femto users to macrocells, a distributed femtocell power control algorithm is developed based on non-cooperative game theory in [7], while in [4] femto users are priced for causing interference to macrocells in the power allocation based on a Stackelberg model. In [24], cross-tier interference is mitigated through both open-loop and closed-loop uplink power control. In [16], a distributed power control scheme is proposed based on a supermodular game.

A lot of work has also been done on subchannel allocation in co-channel deployment of femtocells. In [25], a hybrid frequency assignment scheme is proposed for femtocells deployed within coverage of a macrocell. In [17], distributed channel selection schemes are proposed for femtocells to avoid inter-cell interference, at the cost of reduced frequency reuse efficiency. In [18], a subchannel allocation algorithm based on a potential game model is proposed to mitigate both co-tier and cross-tier interference.

Recently, several studies considering both power and subchannel allocation in femtocells have been reported. In [26], a joint power and subchannel allocation algorithm is proposed to maximize the total capacity of densely deployed femtocells, but neither the interference caused by femtocells to macrocells nor the fairness between femto users has been considered. In the collaborative resource allocation scheme proposed in [29], cross-tier interference is approximated as additive white Gaussian noise (AWGN). In the Lagrangian dual decomposition based resource allocation scheme [27], constraints on cross-tier interference are used in power allocation, but subchannels are assigned randomly to femto users. In [28], a distributed downlink resource allocation scheme based on a potential game and convex optimization is proposed to increase the total capacity of macrocells and femtocells, but at the price of reduced femtocell capacity. In [21], the distributed power and subchannel allocation for co-channel deployed femtocells is modeled as a non-cooperative game, for which a Nash Equilibrium is obtained based on a time-sharing subchannel allocation, but the constraint on maximum femto-user transmit power is ignored in solving the non-cooperative game.

Game theory has been considered to mitigate interference in two-tier networks with co-channel deployed femtocells. In [4, 21], the minimization of co-tier and cross-tier interference though power control based on game theory is investigated. In [7], the authors introduce a distributed utility-based SINR adaptation algorithm in order to alleviate cross-tier interference caused by co-channel femtocells to the macrocell. In [19], a decentralized femtocell access strategy based on non-cooperative game is proposed to manage the interference between nearby femtocells and from femtocells to macrocells. The authors in [4] propose a distributed power control algorithm for spectrum-sharing femtocell network using Stackelberg game, which is very effective in distributed power allocation and macrocell protection while requiring minimal network overhead.

Recently, several studies considering pricing techniques together with power controls have been reported. In [21], the distributed cross-tier interference pricing in power allocation for co-channel deployed femtocells is modeled as a non-cooperative game, but the constraint on maximum femto-user transmit power is ignored in solving the non-cooperative game. For alleviating uplink interference caused by co-channel femto users to macrocells, a distributed femtocell power control algorithm is developed based on non-cooperative game theory in [7]; while in [4] femto users are priced for causing interference to macrocells in the power allocation based on a Stackelberg model.

1.3   Challenges and Issues

Interference mitigation based on resource allocation has been widely analyzed to maintain user's QoS, e.g., signal to interference and noise ratio (SINR) capacity, while alleviating cross-tier interference in two-tier networks. In [7], non-cooperative power allocation with SINR adaptation is used to alleviate the uplink interference suffered by macrocells; while in [4], Stackelberg game based power control is formulated to maximize femtocells' total capacity under cross-tier interference constraints. However, subchannel allocation is not considered. In [26], a joint subchannel and power allocation algorithm is proposed to maximize total capacity in dense femtocell deployments. While in [27], a Lagrangian dual decomposition based resource allocation scheme with constraints on cross-tier interference in power allocations is used. In [21], the distributed subchannel and power allocation for co-channel deployed femtocells is modeled as a non-cooperative game, for which a Nash Equilibrium is obtained based on a time-sharing subchannel allocation scheme. However, in these works, joint subchannel and power allocation with considerations of users' QoS and cross-tier interference is not studied. In [20], a distributed modulation and coding scheme in conjunction with subchannel and power allocation that supports different throughput constraints per users is proposed, but it does not consider two-tier networks. There have been very few works in the literature that make efforts to maximize the capacity of a two-tier network while jointly considering cross-tier interference, QoS requirements and the fairness among users in femtocell networks.

Moreover, femtocell networks should support the heterogeneous QoS requirements of delay sensitive services such as online gaming and video phone calls, while maximizing the throughput of delay tolerant services [32]. However, to the best of our knowledge, resource allocation for heterogeneous QoS users in femtocells has not been studied in previous works. Resource allocation strategies that have been widely studied in spectrum underlay Cognitive Radio (CR) networks [35,36] cannot be directly applied for Interference mitigation in two-tier macrocell and femtocell networks [4].

References

1. D. López-Pérez, A. Valcarce, G. de la Roche, and J. Zhang, "Ofdma femtocells: A roadmap on interference avoidance," *IEEE Commun. Mag.*, vol. 47, no. 9, pp. 41–48, 2009.
2. V. Chandrasekhar and J. G. Andrews, "Femtocell networks: A survey," *IEEE Commun. Mag.*, vol. 46, no. 9, pp. 59–67, 2008.
3. *E-UTRA and E-UTRAN Overall Description*, 3GPP Std. TS 36.300 v10.0.0, 2010.
4. X. Kang, R. Zhang, and M. Motani, "Price-based resource allocation for spectrum-sharing femtocell networks: a stackelberg game approach," *IEEE J. Sel. Areas in Commun.*, 2012.
5. R. Madan, A. Sampath, N. Bhushan, A. Khandekar, J. Borran, and T. Ji, "Impact of coordination delay on distributed scheduling in lte-a femtocell networks," in *Global Telecommunications Conference (GLOBECOM 2010), 2010 IEEE*, 2010, pp. 1–5.

6. K.-S. Lee and D.-H. Cho, "Cooperation based resource allocation for improving inter-cell fairness in femtocell systems," in *Personal Indoor and Mobile Radio Communications (PIMRC), 2010 IEEE 21st International Symposium on*, 2010, pp. 1168–1172.

7. V. Chandrasekhar, J. G. Andrews, T. Muharemovic, Z. Shen, and A. Gatherer, "Power control in two-tier femtocell networks," *IEEE Trans. Wireless Commun.*, vol. 8, no. 8, pp. 4316–4328, 2009.

8. J. Zhang, Z. Zhang, K. Wu, andA. Huang, "Optimal distributed subchannel, rate and power allocation algorithm in ofdm-based two-tier femtocell networks," in *IEEE VTC'10 Spring*, May, pp. 1–5.

9. M. Dorigo, V. Maniezzo, and A. Colorni, "Ant system: optimization by a colony of cooperating agents," *Systems, Man, and Cybernetics, Part B: Cybernetics, IEEE Transactions on*, vol. 26, no. 1, pp. 29–41, 1996.

10. Y. Zhao, X. Xu, Z. Hao, X. Tao, and P. Zhang, "Resource allocation in multiuser ofdm system based on ant colony optimization," in *Wireless Communications and Networking Conference (WCNC), 2010 IEEE*, 2010, pp. 1–6.

11. R. Lin, K. Niu, W. Xu, and Z. He, "A two-level distributed sub-carrier allocation algorithm based on ant colony optimization in ofdma systems," in *Vehicular Technology Conference (VTC 2010-Spring), 2010 IEEE 71st*, 2010, pp. 1–5.

12. X. Zhang, W. Ye, S. Feng, and H. Zhuang, "Adaptive resource allocation for ofdma system based on ant colony algorithm," in *Information Science and Engineering (ICISE), 2009 1st International Conference on*, 2009, pp. 2526–2529.

13. D. Wu and R. Negi, "Effective capacity: a wireless link model for support of quality of service," *Wireless Communications, IEEE Transactions on*, vol. 2, no. 4, pp. 630–643, 2003.

14. T. Stutzle and M. Dorigo, "A short convergence proof for a class of ant colony optimization algorithms," *Evolutionary Computation, IEEE Transactions on*, vol. 6, no. 4, pp. 358–365, 2002.

15. R. Jain, D.-M. Chiu, and W. R. Hawe, *A quantitative measure of fairness and discrimination for resource allocation in shared computer system.* Eastern Research Laboratory, Digital Equipment Corporation, 1984.

16. E. J. Hong, S. Y. Yun, and D.-H. Cho, "Decentralized power control scheme in femtocell networks : A game theoretic approach," in *IEEE PIMRC'09*, pp. 1–5.

17. C. Lee, J.-H. Huang, and L.-C. Wang, "Distributed channel selection principles for femtocells with two-tier interference," in *Vehicular Technology Conference (VTC 2010-Spring), 2010 IEEE 71st*, may 2010, pp. 1–5.

18. I. Mustika, K. Yamamoto, H. Murata, and S. Yoshida, "Potential game approach for self-organized interference management in closed access femtocell networks," in *Vehicular Technology Conference (VTC Spring), 2011 IEEE 73rd*, may 2011, pp. 1–5.

19. S. Barbarossa, S. Sardellitti, A. Carfagna, and P. Vec-chiarelli, "Decentralized interference management in femtocells: A game-theoretic approach," in *IEEE CROWNCOM'10*, June 2010, pp. 1–5.

20. D. Lopez-Perez, A. Ladanyi, A. Juttner, H. Rivano, and J. Zhang, "Optimization method for the joint allocation of modulation schemes, coding rates, resource blocks and power in self-organizing lte networks," in *INFOCOM, 2011 Proceedings IEEE*, april 2011, pp. 111–115.

21. J.-H. Yun and K. G. Shin, "Adaptive interference management of ofdma femtocells for co-channel deployment," *IEEE J. Sel. Areas in Commun.*, vol. 29, no. 6, pp. 1225–1241, 2011.

22. K. Son, S. Lee, Y. Yi, and S. Chong, "Refim: A practical interference management in heterogeneous wireless access networks," *IEEE J. Sel. Areas in Commun.*, vol. 29, no. 6, pp. 1260–1272, 2011.

23. M. Yavuz, F. Meshkati, S. Nanda, A. Pokhariyal, N. Johnson, B. Raghothaman, and A. Richardson, "Interference management and performance analysis of umts/hspa+ femtocells," *IEEE Commun. Mag.*, vol. 47, no. 9, pp. 102–109, Sep. 2009.

24. H.-S. Jo, C. Mun, J. Moon, and J.-G. Yook, "Interference mitigation using uplink power control for two-tier femtocell networks," *IEEE Trans. Wireless Commun.*, vol. 8, no. 10, pp. 4906–4910, Oct. 2009.

25. I. Guvenc, M.-R. Jeong, F. Watanabe, and H. Inamura, "A hybrid frequency assignment for femtocells and coverage area analysis for co-channel operation," *IEEE Commun. Lett.*, vol. 12, no. 12, pp. 880–882, Dec. 2008.
26. J. Kim and D.-H. Cho, "A joint power and subchannel allocation scheme maximizing system capacity in indoor dense mobile communication systems," *IEEE Trans. Veh. Technol.*, vol. 59, no. 9, pp. 4340–4353, 2010.
27. J. Zhang, Z. Zhang, K. Wu, and A. Huang, "Optimal distributed subchannel, rate and power allocation algorithm in ofdm-based two-tier femtocell networks," in *Proc. Veh. Technol. Conf.*, May 2010, pp. 1–5.
28. L. Giupponi and C. Ibars, "Distributed interference control in ofdma-based femtocells," in *IEEE PIMRC'10*, Sept. 2010, pp. 1201–1206.
29. K. Lee, H. Lee, and D.-H. Cho, "Collaborative resource allocation for self-healing in self-organizing networks," in *IEEE Int. Conf. Commun.*, June 2011, pp. 1–5.
30. J. W. Huang and V. Krishnamurthy, "Cognitive base stations in lte/3gpp femtocells: A correlated equilibrium game-theoretic approach," *IEEE Trans. Wireless Commun.*, vol. 59, no. 12, pp. 3485–3493, Dec. 2011.
31. D.-C. Oh, H.-C. Lee, and Y.-H. Lee, "Power control and beamforming for femtocells in the presence of channel uncertainty," *IEEE Trans. Veh. Technol.*, vol. 60, no. 6, pp. 2545–2554, July 2011.
32. M. Tao, Y.-C. Liang, and F. Zhang, "Resource allocation for delay differentiated traffic in multiuser ofdm systems," *IEEE Trans. Wireless Commun.*, vol. 7, no. 6, pp. 2190–2201, June 2008.
33. Y. Zhang and C. Leung, "Cross-layer resource allocation for mixed services in multiuser ofdm-based cognitive radio systems," *IEEE Trans. Veh. Technol.*, vol. 58, no. 8, pp. 4605–4619, Oct. 2009.
34. S. Haykin, "Cognitive radio: brain-empowered wireless communications," *IEEE J. Sel. Areas in Commun.*, vol. 23, no. 2, pp. 201–220, Feb. 2005.
35. D. T. Ngo and T. Le-Ngoc, "Distributed resource allocation for cognitive radio networks with spectrum-sharing constraints," *IEEE Trans. Veh. Technol.*, vol. 60, no. 7, pp. 3436–3449, Sept. 2011.
36. R. Xie, F. R. Yu, and H. Ji, "Dynamic resource allocation for heterogeneous services in cognitive radio networks with imperfect channel sensing," *IEEE Trans. Veh. Technol.*, no. 99, p. 1, 2011.

Chapter 2
Ant Colony Algorithm (ACA) Based Downlink Resource Allocation in Femtocells

Abstract This chapter focuses on the resource allocation of femtocells in the Orthogonal Frequency Division Multiple Access (OFDMA) networks. A typical algorithm of swarm intelligence called Ant Colony Optimization (ACO) is adopted to resolve the optimization problem of maximizing the total capacity of femtocells considering the quality of service (QoS) requirement. An ACO based system model for the resource allocation, as well as three different schemes (ACOMAX, ACOPF and ACOCF) that are based on meta-heuristic methods is proposed. Due to the unique characteristics of ACO's heuristic searching mechanism, the proposed algorithms can guarantee a fast convergence speed. Simulation results show that ACOMAX can significantly increase the throughput of the system, and ACOCF as well as ACOPF can satisfy the requirements of throughput and guarantee fairness simultaneously.

2.1 Introduction

Femtocells have been proposed for improving the performance of indoor users [1]. Femtocells are usually comprised of small size, low-power, low-cost, and short-range home base stations. They work in the licensed frequency bands, and are connected to broadband Internet backhaul. As femtocells can meet customer's demands and indoor coverage requirements well, they have been widely introduced in many wireless communication standards, such as WiMAX, and LTE/LTE-Advanced [2]. Therefore, two-tier OFDMA networks comprising macrocells and femtocells are widely expected to improve the coverage and capacity of cellular networks.

In practice, there are still some technical challenges that need to be further addressed before widespread deployment of femtocells. A two-tier network is usually implemented by sharing frequency rather than splitting frequency between tiers [3]. Hence cross-tier interference (CTI) and intra-tier interference (ITI) are the key issues in two-tier networks [4], and maximization of the total data rate

H. Zhang et al., *4G Femtocells: Resource Allocation and Interference Management*,
SpringerBriefs in Computer Science, DOI 10.1007/978-1-4614-9080-7_2,
© The Author(s) 2013

of femtocells with the consideration of CTI and ITI is a hot research area. The related works on femtocell networks in the literature are described in the following. Resource allocation algorithms aiming at the inter-cell interference management in femtocell networks are discussed and evaluated in [5]. A resource allocation scheme considering inter-femtocell fairness is proposed in [6], and in [7] a CTI mitigation algorithm based on power control is developed. The authors in [23] propose a distributed resource allocation algorithm based on Lagrangian dual method. However, with the consideration of CTI, quality of service (QoS) requirements and the fairness among users in femtocell networks, there is little work has been done related to maximize the capacity of the two-tier network.

Ant Colony Algorithm (ACA) is a typical swarm intelligence algorithm [9], which has been used for resource allocation in OFDM systems [10, 11]. The feature of robustness and parallel heuristic in ACA is fit for finding proper solution for resource allocation.But this approach has not been sufficiently explored in the two-tier network literature.

In this chapter, we consider a system model based on ACA in a two-tier network, where femtocells are deployed densely. Our goal is to maximize the total downlink rate of all femtocell users, while considering CTI, QoS requirement and fairness among users. We propose an ACA based algorithms to optimize the sub-channels allocation problem [13], and the performance of these algorithms is evaluated by simulation. Compared with the traditional round robin (RR) algorithm [12], the ACA based algorithm achieves better performance.

The rest of the chapter is organized as follows: Sect. 2.2 introduces the system model and problem formulation. In Sect. 2.3, the resource allocation algorithms based on ACA is presented. Simulation results and performance analysis are provided in Sect. 2.4. Finally, we conclude the chapter in Sect. 2.5.

2.2 System Model and Problem Formulation

2.2.1 System Model

In this chapter, we consider a two-tier OFDMA network as shown in Fig. 2.1, where femtocells are deployed densely [7]. A macrocell base station (MBS) locates in the center of a disc area with a radius of R_m. At a distance Df from the MBS and within the coverage region of the MBS, femtocell base stations (FBS) ($\{B_i\}(i = 1 \cdots K)$) are arranged in a square grid of area D_{grid}^2 sq.km with $\sqrt{K}$ femtocells per dimension, at a distance D_f from MBS. The radius of each femtocell is R_f. Let D denote the distance between a transmitting mobile and the MBS. All femtocells are assumed to be closed access, and femtocells use the same frequency resource that the macrocell uses.

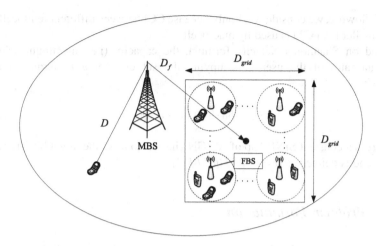

Fig. 2.1 The topology of a two-tier network

It is assumed that there are wired connections for the FBSs to communicate with the MBS. And there is one scheduled active user during each signaling slot in each femtocell. Let $k \in \{1,\ldots,K\}$ denote the scheduled active user connected to its FBS B_i. The system has a total bandwidth B and divides it into N sub-channels, each with a bandwidth of $B_0 = B/N$. The channel fading of each subcarrier is assumed the same within a sub-channel, but may vary cross different sub-channels.

We define $p_{k,n}^{(m)}$ as the transmission power of the MBS on sub-channel n to one of its users, $p_{k,n}^{(i)}$ and $p_{k,n}^{(i')}$ are the transmission powers of the serving FBS i to user k and the neighbor FBS i' to its scheduled user on sub-channel n respectively. Let $g_{k,n}^i$, $g_{k',n}^{(i')}$, and $g_{m,n}^m$ denote the channel gain from the serving FBS i, the interfering FBS i' and the MBS on sub-channel n to user k of femtocell i, respectively. We consider the channel model consisting of large scale fading and Rayleigh fading [14]. Therefore, the received signal to interference and noise ratio (SINR) for user k in femtocell i occupying the sub-channel n is given by:

$$\gamma_{k,n} = \frac{p_{k,n}^i \cdot g_{k,n}^i}{\displaystyle\sum_{k'=1,k'\neq k}^{K} p_{k',n}^{i'} \cdot g_{k',n}^{(i')} + p_{k,n}^{(m)} \cdot g_{m,n}^m + N_0 \cdot B_0} \tag{2.1}$$

where $\displaystyle\sum_{k'=1,k'\neq k}^{K} p_{k',n}^{i'} \cdot g_{k',n}^{(i')}$ is the interference caused by other co-channel femtocells, that is co-channel interference. $p_{k,n}^{(m)} \cdot g_{m,n}^m$ is the interference caused by the macro-cell, and N_0 is the additive white Gaussian noise (AWGN) power spectral density.

In the following, we consider CTI only because CCI between different femtocells is much smaller than CTI caused by macrocell.

Based on Shannon's capacity formula, the capacity (i.e., maximum achievable data rate) of the user k occupying the sub-channel n in femtocell i is given by:

$$r_{k,n} = B_0 \cdot \log_2 \left(1 + \alpha \cdot \gamma_{k,n}\right) \tag{2.2}$$

where α is a constant SINR gap of AWGN channel to meet the target bit error rate (BER), and is defined as $\alpha = -1.5/\ln(5BER)$.

2.2.2   Problem Formulation

Our target is to maximize the total data rate of all users of a femtocell (hereafter we omit the femtocell index i for simplicity), that is:

$$\max \sum_{k=1}^{K} \sum_{n=1}^{N} c_{k,n} r_{k,n}$$

$$s.t. \quad \sum_{k=1}^{K} \sum_{n=1}^{N} c_{k,n} p_{k,n} \le p_{\max}$$

$$p_{k,n} \ge 0, \forall k \in \{1,2,\ldots,K\}, \forall n \in \{1,2,\ldots,N\}$$

$$BER_k \le BER_{request}; \quad c_{k,n} \in \{0,1\}, \quad \forall\, k,\, n \tag{2.3}$$

where $c_{k,n}$ indicates whether sub-channel n is occupied by user k, and is denoted as follows:

$$c_{k,n} = \begin{cases} 1 & \text{if subchannel } n \text{ is occupied by user } k \\ 0 & \text{otherwise} \end{cases} \tag{2.4}$$

the total transmission power of an FBS is constrained by $p_{\max}$, and the power allocated on each sub-channel is nonnegative. The BER of each femtocell user k is upper bounded by the limit $BER_{request}$, which is set according to user's QoS requirement defines a sub-channel can only be used by one user at a time in each cell.

2.3   Resource Allocation Using ACA

In this section, ACA based resource allocation schemes are proposed in femtocells co-existing with a macrocell. We first present a scenario that apply ACA in the resource allocation process for system, and then propose algorithms based on ACA. The algorithms allocate the sub-channels to users under the consideration of cross-tier interference and fairness among users.

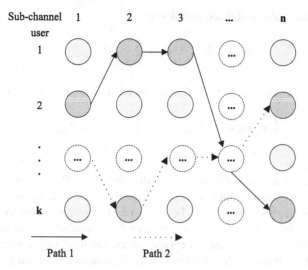

Fig. 2.2 Application of ACA for sub-channel allocation

2.3.1   Application of ACA in Resource Allocation

Traveling Salesman Problem (TSP) [9] is chosen as the basic model for resource allocation. As shown in Fig. 2.2, each node represents a state in the process of solution building by ants. Each node in the kth row denotes active user k ($k \in 1, 2, \ldots, K$). Each node in the nth column denotes sub-channel n and ($n \in 1, 2, \ldots, N$). It is assumed that all the ants depart from the first unoccupied sub-channel at the beginning. The selected routes (in terms of passed nodes) are recorded in a matrix. When node is selected more times, more guideline information will be provided and ant would like to choose it next time. The probability of ant m on sub-channel n choosing user k is given by

$$p_{k,n}^m = \begin{cases} \tau_{k,n}^\alpha \cdot \eta_{k,n}^\beta / (\sum_{l \in N_n^m} \tau_{l,n}^\alpha \cdot \eta_{l,n}^\beta) & if \ \ k \in K \\ 0 & otherwise \end{cases} \qquad (2.5)$$

where $\tau_{k,n}$ and $\eta_{k,n}$ denote the pheromone trail and the heuristic information that ant m on sub-channel n chooses user k, respectively [10]. α and β are the parameters that determine the relative importance of pheromone trail versus heuristic information. If α is bigger, the solution process will stop at a route which likely a good solution; if β is bigger, it is likely that the system will find a good solution but the convergence time may be long. So a proper setting of α and β is important for a good trade-off between the convergence time and system performance. N_n^m denotes the set of available users for ant m on sub-channel n to traverse, which can be updated after each user has been selected.

The heuristic value is defined as follows:

$$\eta_{k,n} = r_{k,n} / (\sum_{l \in N_n^m} r_{l,n}) \tag{2.6}$$

Then ant m on sub-channel n will select user k with the following probability

$$\begin{cases} \arg\max_{l \in N_n^m} \left\{ \tau_{l,n} [\eta_{l,n}]^\beta \right\}, & if \ \ q \le q_0 \\ p_{k,n} & else \end{cases} \tag{2.7}$$

where q is a random variable uniformly distributed in [0,1], $q_0 (0 \le q_0 \le 1)$ is an adjusting coefficient, which means the probability of an ant choosing the best possible node at the current time. Equation (2.7) provides a way to control the convergence rate in acquiring the solution of the optimization problem in (2.3). In general, ants are more likely to choose nodes that have larger pheromone and heuristic values. When each ant arrives at the destination, the local pheromone is updated according to

$$\tau_{k,n} \leftarrow (1 - \rho) \tau_{k,n} + \rho \tau_0 \tag{2.8}$$

where $0 < \rho < 1$ is a variable that controls the evaporation of local pheromone, and can make the system forget the bad route and quickly converge to a good solution; τ_0 is the initial value of local pheromone. The local updating changes pheromone on each node and inspires the selection of nodes whose pheromones change dynamically. When all the ants have traversed all the sub-channels, then the global pheromone updates according to

$$\tau_{k,n} \leftarrow (1 - \theta) \tau_{k,n} + \theta \Delta \tau_{k,n}^c \tag{2.9}$$

where θ is the pheromone evaporation parameter, $0 < \theta < 1$, and $\Delta \tau_{k,n}^c$ denotes the pheromone on the best path that an ant chooses, which is given by

$$\Delta \tau_{k,n}^c = Q / L_{gb} \tag{2.10}$$

where L_{gb} is the global optimal path. Q is a positive coefficient of $\Delta \tau_{k,n}^c$, and the pheromone concentration will be accelerated if Q is bigger. Pheromone evaporation and receiving reinforcement are carried out on the nodes of the best path, which makes the process of searching for the best solution more efficient and direct.

2.3.2   Algorithm Description

In this section, we present different dynamic resource allocation schemes based on ACA, including ACA based Maximizing Capacity with Modified Fairness (ACA-MF), ACA based Maximizing Capacity with Further Fairness (ACA-FF), and ACA based Maximizing Capacity (ACA-MC).

The algorithm ACA-MF is proposed to maintain proportional fairness while increasing overall capacity. ACA-MF first assigns the corresponding optimal sub-channels to each user, and allocates, the remaining sub-channels following a greedy policy. In practical systems, the underlying premise is that a rough proportional fairness is achievable, the throughput is improved and the computational complexity is low. The pseudo code of ACA-MF is given as Algorithm 1.

Algorithm 1 ACA-MF

1: ACA-MF parameters initialization: the initial value of pheromone on each node is set as τ_0; heuristic information is set according to (2.6);
2: sort the sub-channels in decreasing order according to the channel gain $g_{k,n}$;
3: set the number of sub-channels N_k that will be allocated to user k in the beginning, here we design $N_k = K$ for any k, and assign these sub-channels to each user in the following way.
4: **for** $k = 1$ to K **do**
5: $\quad n = \arg\max_{n' \in N} \left| g_{k,n'} \right|$
$\quad c_{k,n} = 1, N_k = N_k - 1, N = N \backslash \{n\}$
6: **end for**
7: for the rest sub-channels, each ant starts from the first unoccupied sub-channel to traverse, and will select users for the current sub-channel according to (2.7), where the pheromone and heuristic information are the subset from step 1 on the rest numbers in N;
8: carry out the local pheromone updating for each sub-channel according to (2.8);
9: carry out the global pheromone updating when complete solutions are acquired according to (2.9);
10: when it achieves the maximum iteration number or convergence, the algorithm ends; else go to step 1, calculate (16);

Compared with ACA-MF, ACA-FF provides improved proportional fairness among users, but has a lower capacity. The pseudo code of ACA-FF is given as Algorithm 2.

Algorithm 2 ACA-FF

1: Starting from the first unoccupied sub-channel, ants select users for the current sub-channel according to

$$\eta_{k,n} = \frac{\log(1 + r_{k,n} + R_k) - \log(1 + R_k)}{\sum\limits_{l \in N_n^m} (\log(1 + r_{k,n} + R_l) - \log(1 + R_l))}$$

where the heuristic value is defined as in [10], R_k indicates the transmission rate of user k at current time.
2: the local pheromone updates according to (2.8);
3: the global pheromone updates according to (2.9);
4: if the algorithm achieves the maximum number of iteration or convergence, the algorithm ends; else go to step 1.

ACA-MC consists of only steps 4–7 of Algorithm 1, aiming to maximize the capacity of the system.

2.3.3 Parameters and Convergence of ACA

The parameter τ_0 is the initial setting of pheromone trail. If τ_0 is too small, the search will end up with a local-best solution; otherwise, many iterations at the beginning in the starting will be wasted. The parameter $\eta_{k,n}$ plays a guiding role in ants' search for solution at the beginning when the values of pheromone are set the same for all nodes. The definition of heuristic information influences the solution process and the fairness among users. The function of parameters α and β has been illustrated in Sect. 2.3.1.

The convergence of ACA has been proved in [15]. With the use of parallel processors, the convergence time can be reduced significantly.

2.4 Performance Evaluation

In order to evaluate the performance of the algorithms based on ACA, Monte-Carlo simulation is used. The simulation parameters are summarized in Table 2.1. System capacity, fairness among users and the throughput improvement of the presented schemes are evaluated and compared with RR algorithm. We consider the system model described in Sect. 2.2. The number of femtocells is chosen from the set $F_n = \{4, 9, 16, 25, 36, 49\}$.

Table 2.1 System parameters used in simulation

Parameter	Value
Macrocell/femtocell radius	288/5 m
Grid size D_{grid}	100 m
Carrier frequency $f_{c,MHz}$	2 GHz
System bandwidth B	10 MHz
MBS/ FBS TX power	46/20 dBm
Channel model	Large scale Fading and Rayleigh
Number of the sub-channels	50
Thermal noise density	174 dBm/Hz
Out-/In-door path loss exponent	4/3
ACA parameters	$\tau_0 = 1, \alpha = 1, \beta = 3$
Monte-Carlo times	100

Figure 2.3 shows the total capacity of the femtocells versus the number of femtocells. As can be seen from the figure, the proposed ACA-MF algorithm improves the total capacity of the femtocells over the RR algorithm significantly. Apparently, ACA-MC achieves the best throughput among the four algorithms for it

Fig. 2.3 Capacity
comparison of different
algorithms

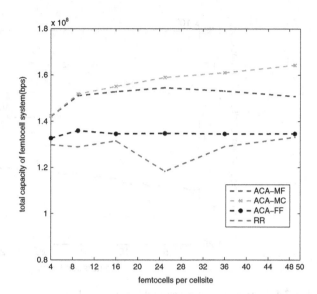

always distributes the optimal sub-channels to the users, followed by ACA-MF. The capacity that RR achieves is the most unstable, because of the mutual independence of the sub-channels when allocated to the users in a round robin fashion. It can also be seen that in ACA-MF and ACA-FF algorithms, the total capacity of femtocells decreases a little when the number of femtocells is larger than 25. This is mainly because the interference caused by users is too much while the number of iterations is insufficient. The capacity obtained by ACA-MC always increases with the number of femtocells because it always allocate the optimal sub-channels to users.

Figure 2.4 illustrates the capacity improvements of ACA algorithms when the number of femtocells is 9. It can be seen that the ACA based algorithms can converge within a limited number of iterations, and all the ACA based algorithms improve the capacity greatly versus RR. In the early iterations, ACA-MF scheme can obtain higher capacity than ACA-MC, because it distributes the optimal sub-channels to each user at the beginning when the ACA-MC hasn't found the global optimal solution yet. ACA-MF achieves a good solution a little later than ACA-MC, but still holds a higher capacity than ACA-FF and RR.

To compare the user-fairness performance of the four considered schemes, we use the Raj Jain Fairness Index [16] to measure the fairness among users, which is defined as:

$$F(r_1, r_2, \ldots, r_K) = (\sum_{k=1}^{K} r_k)^2 / (K \cdot \sum_{k=1}^{K} r_k^2) \qquad (2.11)$$

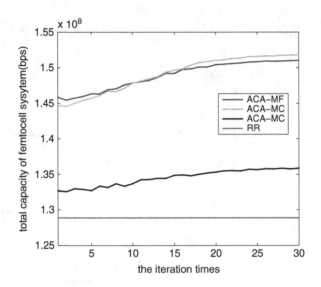

Fig. 2.4 Capacity improvements of different algorithms

Fig. 2.5 Fairness comparison of different algorithms

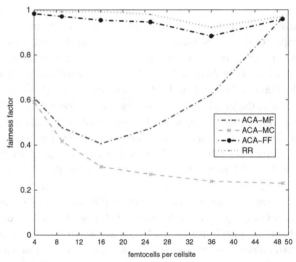

where r_k is the already assigned rate of the corresponding user k, and K is the total number of femto users. The bigger the fairness index, the better the fairness among users.

As shown in Fig. 2.5, the conventional RR provides the best fairness because it minimizes the variance of sub-channel resources allocated to different users. However, in practice, there's no need for users to occupy similar amounts of sub-channels. The ACA-FF algorithm provides proportional fairness among users that is close to that of RR but is much better than ACA-MC schemes. The fairness index of ACA-MF first decreases and then increases with the number of femtocells, and reaches a similar value as the RR and ACO-FF schemes when the number of femtocells is 49.

2.5 Conclusion

Sub-channel allocation algorithms based on Ant Colony Algorithm are presented in this chapter. As a typical meta-heuristic method, ACA provides simple and robust way for resource-allocation optimization. We formulate the resource allocation problem as path searching in a graph, and use the pheromone trail and the heuristic information to guide the construction of solution construction. In comparison with the traditional RR algorithm, better throughput can be achieved by ACA based algorithms. ACA-MF and ACA-FF can guarantee the fairness among users while meeting rate requirements.

Acknowledgements The authors would like to thank Ms. Deli Liu for her contribution and helpful discussions. This research has been supported by National Key Technology R&D Program of China (2010ZX03003-001-01, 2011ZX03003-002-01), National Natural Science Foundation of China (61101109), Co-building Project of Beijing Municipal Education Commission "G-RAN based Experimental Platform for Future Mobile Communications", "Research on Resource Allocation and Scheduling Strategy of Future Wireless Communication System" and "Cooperative Communications Platform for Multi-agent Multimedia Communications", Key Fund of Beijing Key Laboratory on Future Network Research.

References

1. V. Chandrasekhar and J. G. Andrews, "Femtocell networks: A survey," *IEEE Commun. Mag.*, vol. 46, no. 9, pp. 59–67, 2008.
2. *E-UTRA and E-UTRAN Overall Description*, 3GPP Std. TS 36.300 v10.0.0, 2010.
3. X. Kang, R. Zhang, and M. Motani, "Price-based resource allocation for spectrum-sharing femtocell networks: a stackelberg game approach," *IEEE J. Sel. Areas in Commun.*, 2012.
4. D. López-Pérez, A. Valcarce, G. de la Roche, and J. Zhang, "Ofdma femtocells: A roadmap on interference avoidance," *IEEE Commun. Mag.*, vol. 47, no. 9, pp. 41–48, 2009.
5. R. Madan, A. Sampath, N. Bhushan, A. Khandekar, J. Borran, and T. Ji, "Impact of coordination delay on distributed scheduling in lte-a femtocell networks," in *Global Telecommunications Conference (GLOBECOM 2010), 2010 IEEE*, 2010, pp. 1–5.
6. K.-S. Lee and D.-H. Cho, "Cooperation based resource allocation for improving inter-cell fairness in femtocell systems," in *Personal Indoor and Mobile Radio Communications (PIMRC), 2010 IEEE 21st International Symposium on*, 2010, pp. 1168–1172.
7. V. Chandrasekhar, J. G. Andrews, T. Muharemovic, Z. Shen, and A. Gatherer, "Power control in two-tier femtocell networks," *IEEE Trans. Wireless Commun.*, vol. 8, no. 8, pp. 4316–4328, 2009.
8. J. Zhang, Z. Zhang, K. Wu, and A. Huang, "Optimaldistributed subchannel, rate and power allocation algorithm in ofdm-based two-tier femtocell networks," in *IEEE VTC'10 Spring*, May, pp. 1–5.

9. M. Dorigo, V. Maniezzo, and A. Colorni, "Ant system: optimization by a colony of cooperating agents," *Systems, Man, and Cybernetics, Part B: Cybernetics, IEEE Transactions on*, vol. 26, no. 1, pp. 29–41, 1996.

10. Y. Zhao, X. Xu, Z. Hao, X. Tao, and P. Zhang, "Resource allocation in multiuser ofdm system based on ant colony optimization," in *Wireless Communications and Networking Conference (WCNC), 2010 IEEE*, 2010, pp. 1–6.

11. R. Lin, K. Niu, W. Xu, and Z. He, "A two-level distributed sub-carrier allocation algorithm based on ant colony optimization in ofdma systems," in *Vehicular Technology Conference (VTC 2010-Spring), 2010 IEEE 71st*, 2010, pp. 1–5.

12. X. Zhang, W. Ye, S. Feng, and H. Zhuang, "Adaptive resource allocation for ofdma system based on ant colony algorithm," in *Information Science and Engineering (ICISE), 2009 1st International Conference on*, 2009, pp. 2526–2529.

13. D. Liu, H. Zhang, W. Zheng, X. Wen, L. Li, "The Sub-channel Allocation Algorithm in Femtocell Networks Based on Ant Colony Optimization," accepted by MILCOM 2012.

14. D. Wu and R. Negi, "Effective capacity: a wireless link model for support of quality of service," *Wireless Communications, IEEE Transactions on*, vol. 2, no. 4, pp. 630–643, 2003.

15. T. Stutzle and M. Dorigo, "A short convergence proof for a class of ant colony optimization algorithms," *Evolutionary Computation, IEEE Transactions on*, vol. 6, no. 4, pp. 358–365, 2002.

16. R. Jain, D.-M. Chiu, and W. R. Hawe, *A quantitative measure of fairness and discrimination for resource allocation in shared computer system.*   Eastern Research Laboratory, Digital Equipment Corporation, 1984.

17. J.-H. Yun and K. G. Shin, "Adaptive interference management of ofdma femtocells for co-channel deployment," *IEEE J. Sel. Areas in Commun.*, vol. 29, no. 6, pp. 1225–1241, 2011.

18. K. Son, S. Lee, Y. Yi, and S. Chong, "Refim: A practical interference management in heterogeneous wireless access networks," *IEEE J. Sel. Areas in Commun.*, vol. 29, no. 6, pp. 1260–1272, 2011.

19. M. Yavuz, F. Meshkati, S. Nanda, A. Pokhariyal, N. Johnson, B. Raghothaman, and A. Richardson, "Interference management and performance analysis of umts/hspa+ femtocells," *IEEE Commun. Mag.*, vol. 47, no. 9, pp. 102–109, Sep. 2009.

20. H.-S. Jo, C. Mun, J. Moon, and J.-G. Yook, "Interference mitigation using uplink power control for two-tier femtocell networks," *IEEE Trans. Wireless Commun.*, vol. 8, no. 10, pp. 4906–4910, Oct. 2009.

21. I. Guvenc, M.-R. Jeong, F. Watanabe, and H. Inamura, "A hybrid frequency assignment for femtocells and coverage area analysis for co-channel operation," *IEEE Commun. Lett.*, vol. 12, no. 12, pp. 880–882, Dec. 2008.

22. J. Kim and D.-H. Cho, "A joint power and subchannel allocation scheme maximizing system capacity in indoor dense mobile communication systems," *IEEE Trans. Veh. Technol.*, vol. 59, no. 9, pp. 4340–4353, 2010.

23. J. Zhang, Z. Zhang, K. Wu, and A. Huang, "Optimal distributed subchannel, rate and power allocation algorithm in ofdm-based two-tier femtocell networks," in *Proc. Veh. Technol. Conf.*, May 2010, pp. 1–5.

24. L. Giupponi and C. Ibars, "Distributed interference control in ofdma-based femtocells," in *IEEE PIMRC'10*, Sept. 2010, pp. 1201–1206.

25. K. Lee, H. Lee, and D.-H. Cho, "Collaborative resource allocation for self-healing in self-organizing networks," in *IEEE Int. Conf. Commun.*, June 2011, pp. 1–5.

26. J. W. Huang and V. Krishnamurthy, "Cognitive base stations in lte/3gpp femtocells: A correlated equilibrium game-theoretic approach," *IEEE Trans. Wireless Commun.*, vol. 59, no. 12, pp. 3485–3493, Dec. 2011.

27. D.-C. Oh, H.-C. Lee, and Y.-H. Lee, "Power control and beamforming for femtocells in the presence of channel uncertainty," *IEEE Trans. Veh. Technol.*, vol. 60, no. 6, pp. 2545–2554, July 2011.

28. M. Tao, Y.-C. Liang, and F. Zhang, "Resource allocation for delay differentiated traffic in multiuser ofdm systems," *IEEE Trans. Wireless Commun.*, vol. 7, no. 6, pp. 2190–2201, June 2008.

29. Y. Zhang and C. Leung, "Cross-layer resource allocation for mixed services in multiuser ofdm-based cognitive radio systems," *IEEE Trans. Veh. Technol.*, vol. 58, no. 8, pp. 4605–4619, Oct. 2009.

30. S. Haykin, "Cognitive radio: brain-empowered wireless communications," *IEEE J. Sel. Areas in Commun.*, vol. 23, no. 2, pp. 201–220, Feb. 2005.

31. D. T. Ngo and T. Le-Ngoc, "Distributed resource allocation for cognitive radio networks with spectrum-sharing constraints," *IEEE Trans. Veh. Technol.*, vol. 60, no. 7, pp. 3436–3449, Sept. 2011.

32. R. Xie, F. R. Yu, and H. Ji, "Dynamic resource allocation for heterogeneous services in cognitive radio networks with imperfect channel sensing," *IEEE Trans. Veh. Technol.*, no. 99, p. 1, 2011.

33. K. W. Choi, E. Hossain, and D. I. Kim, "Downlink subchannel and power allocation in multi-cell ofdma cognitive radio networks," *IEEE Trans. Wireless Commun.*, vol. 10, no. 7, pp. 2259–2271, July 2011.

34. Y. Ma, D. I. Kim, and Z. Wu, "Optimization of ofdma-based cellular cognitive radio networks," *IEEE Trans. Commun.*, vol. 58, no. 8, pp. 2265–2276, Aug. 2010.

35. *Way forward proposal for (H)eNB to HeNB mobility*, 3GPP Std. R3-101 849, 2010.

36. H.-S. Jo, C. Mun, J. Moon, and J.-G. Yook, "Interference mitigation using uplink power control for two-tier femtocell networks," *IEEE Trans. Wireless Commun.*, vol. 8, no. 10, pp. 4906–4910, Oct. 2009.

37. S. Boyd and L. Vandenberghe, *Convex Optimization*. Cambridge University Press, 2004.

38. D. W. K. Ng and R. Schober, "Resource allocation and scheduling in multi-cell ofdma systems with decode-and-forward relaying," *IEEE Trans. Wireless Commun.*, vol. 10, no. 7, pp. 2246–2258, July 2011.

39. C. Y. Wong, R. Cheng, K. Lataief, and R. Murch, "Multiuser ofdm with adaptive subcarrier, bit, and power allocation," *IEEE J. Sel. Areas in Commun.*, vol. 17, no. 10, pp. 1747–1758, Oct. 1999.

40. W. Yu and R. Lui, "Dual methods for nonconvex spectrum optimization of multicarrier systems," *IEEE Trans. Commun.*, vol. 54, no. 7, pp. 1310–1322, July 2006.

41. J. K. Chen, G. de Veciana, and T. S. Rappaport, "Site-specific knowledge and interference measurement for improving frequency allocations in wireless networks," *IEEE Trans. Veh. Technol.*, vol. 58, no. 5, pp. 2366–2377, June 2009.

42. *Further Advancements for E-UTRA, Physical Layer Aspects*, 3GPP Std. TR 36.814 v9.0.0, 2010.

43. Z. Shen, J. G. Andrews, and B. L. Evans, "Adaptive resource allocation in multiuser ofdm systems with proportional rate constraints," *IEEE Trans. Wireless Commun.*, vol. 4, no. 6, pp. 2726–2737, Nov. 2005.

Chapter 3
Cross-Tier Interference Pricing Based Uplink Resource Allocation in Two-Tier Networks

Abstract Femtocells have been considered as a promising technology to provide better indoor coverage and spatial reuse gains. However, the co-channel deployment of macrocells and femtocells is still facing challenges arising from potentially severe inter-cell interference. In this paper, we investigate the uplink resource allocation problem of femtocells in co-channel deployment with macrocells. We first model the uplink power and subchannel allocation in femtocells as a non-cooperative game, where inter-cell interference is taken into account in maximizing the femtocell capacity and uplink femto-to-macro interference is alleviated by charging each femto user a price proportional to the interference that it causes to the macrocell. Based on the non-cooperative game, we then devise a semi-distributed algorithm for each femtocell to first assign subchannels to femto users and then allocate power to subchannels. Simulation results show that the proposed interference-aware femtocell uplink resource allocation algorithm is able to provide improved capacities for not only femtocells but also the macrocell, as well as comparable or even better tiered fairness in the two-tier network, as compared with existing unpriced subchannel assignment algorithm and modified iterative water filling based power allocation algorithm.

3.1 Introduction

Nowadays above 50% of voice services and 70% of data traffics occur indoors [1]. Insufficient indoor coverage of macrocells has led to increasing interest in femtocells, which have been considered in major wireless communication standards such as 3GPP LTE/LTE-Advanced [2]. Dedicated-channel deployment of femtocells, where femtocells and macrocells are assigned with different (or orthogonal) frequency bands, may not be preferred by operators due to the scarcity of spectrum resources and difficulties in implementation. While in co-channel deployment, where femtocells and macrocells share the same spectrum, cross-tier interference could be severe [3], especially when femtocell base stations (FBSs) are deployed

H. Zhang et al., *4G Femtocells: Resource Allocation and Interference Management*,
SpringerBriefs in Computer Science, DOI 10.1007/978-1-4614-9080-7_3,
© The Author(s) 2013

close to a macrocell base station (MBS) [5]. Due to the fundamental role of macrocells in providing blanket cellular coverage, their capacities and coverage should not be affected by co-channel deployment of femtocells.

Power control has been widely used to mitigate inter-cell interference in co-channel deployment of femtocells. For alleviating uplink interference caused by co-channel femto users to macrocells, a distributed femtocell power control algorithm is developed based on non-cooperative game theory in [6], while in [7] femto users are priced in power allocation for causing interference to macrocells based on a Stackelberg model. In [8], cross-tier interference is mitigated through both open-loop and closed-loop uplink power control. In [9], a distributed power control scheme is proposed based on a supermodular game.

A lot of work has also been done on subchannel allocation in co-channel deployment of femtocells. In [10], a hybrid frequency assignment scheme is proposed for femtocells deployed within coverage of a macrocell. In [11], distributed channel selection schemes are proposed for femtocells to avoid inter-cell interference, at the cost of reduced frequency reuse efficiency. In [12], a subchannel allocation algorithm based on a potential game model is proposed to mitigate both co-tier and cross-tier interference.

Recently, several studies considering both power and subchannel allocation in femtocells have been reported. In [13], a joint power and subchannel allocation algorithm is proposed to maximize the total capacity of densely deployed femtocells, but the interference caused by femtocells to macrocells is not considered. In the collaborative resource allocation scheme [14], cross-tier interference is approximated as additive white Gaussian noise (AWGN). In the Lagrangian dual decomposition based resource allocation scheme [15], constraints on cross-tier interference are used in power allocation, but subchannels are assigned randomly to femto users. In [16], a distributed downlink resource allocation scheme based on a potential game and convex optimization is proposed to increase the total capacity of macrocells and femtocells, but at the price of reduced femtocell capacity. In [17], the distributed power and subchannel allocation for co-channel deployed femtocells is modeled as a non-cooperative game, for which a Nash Equilibrium is obtained based on a time-sharing subchannel allocation, but the constraint on maximum femto-user transmit power is ignored in solving the non-cooperative game.

In this chapter, we focus on the uplink power and subchannel allocation problem of orthogonal frequency division multiple access (OFDMA) based femtocells in co-channel deployment with macrocells [18, 19]. We first model the uplink power and subchannel allocation in femtocells as a non-cooperative game, where inter-cell interference is taken into account in maximizing femtocell capacity and uplink interference from femto users to the macrocell is alleviated by charging each femto user a price proportional to the amount of interference that it causes to the macrocell. Based on the non-cooperative game, we then devise a semi-distributed algorithm for each femtocell to first assign subchannels to femto users and then allocate power to subchannels accordingly. Simulation comparisons with existing unpriced subchannel assignment and modified iterative water filling (MIWF) based power allocation algorithms show that the proposed interference-aware femtocell uplink resource allocation algorithm is able to provide improved capacities for not only femtocells but also the macrocell, as well as comparable or even better tiered fairness in a co-channel two-tier network.

The rest of this chapter is organized as follows. The system model and problem formulation are presented in Sect. 3.2. In Sect. 3.3, the interference-aware femtocell uplink resource allocation algorithm is proposed. Performance of the proposed algorithm is evaluated by simulations in Sect. 3.4. Finally, Sect. 3.5 concludes the paper.

3.2 System Model and Problem Formulation

3.2.1 System Model

As shown in Fig. 3.1, we consider a two-tier OFDMA network where K co-channel FBSs are randomly overlaid on a macrocell. We focus on resource allocation in the uplink of femtocells, that is, the subchannel assignment to femto users and the power allocation on subchannels in femtocells. Let M and F denote the numbers of active macro users camping on the macrocell and active femto users camping on each femtocell, respectively. Users are uniformly distributed in the coverage area of their serving cell. All femtocells are assumed to be closed access [20]. The OFDMA system has a bandwidth of B, which is divided into N subchannels. Channel fading

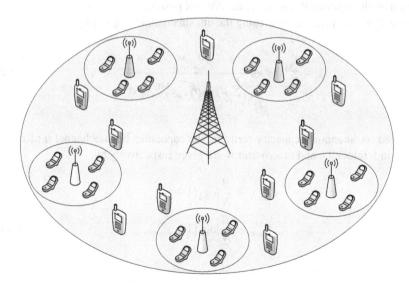

Fig. 3.1 Topology of the two-tier network comprising by a macrocell and K co-channel femtocells

on each subcarrier is assumed the same within a subchannel, but may vary across different subchannels. We assume that channel fading is composed of path loss and frequency-flat Rayleigh fading.

We denote $g_{k,u,n}^{\mathrm{MF}}$ and $g_{j,k,u,n}^{\mathrm{FF}}$ as the channel gains on subchannel n from femto user u in femtocell k to the MBS and FBS j, respectively, where $j,k \in \{1,2,\ldots,K\}$, $u \in \{1,2,\ldots,F\}$, and $n \in \{1,2,\ldots,N\}$; denote $g_{w,n}^{\mathrm{M}}$ and $g_{k,w,n}^{\mathrm{FM}}$ as the channel gains on subchannel n from macro user $w(\in \{1,2,\ldots,M\})$ to the MBS and FBS k, respectively; denote $p_{k,u,n}^{\mathrm{F}}$ and $p_{w,n}^{\mathrm{M}}$ as the transmit power levels on subchannel n of femto user u in femtocell k and macro user w, respectively. Then we define $P_n = [p_{k,u,n}^{\mathrm{F}}]_{K \times F}$ as the power allocation matrix of the K femtocells on subchannel n, and $A_n = [a_{k,u,n}]_{K \times F}$ as the subchannel assignment indication matrix for the K femtocells on subchannel n, where $a_{k,u,n} = 1$ if subchannel n is assigned to femto user u in femtocell k, and $a_{k,u,n} = 0$ otherwise.

The received signal to interference plus noise ratio (SINR) for femto user u on the nth subchannel in the kth femtocell is given by

$$\gamma_{k,u,n}^{\mathrm{F}} = \frac{a_{k,u,n} p_{k,u,n}^{\mathrm{F}} g_{k,k,u,n}^{\mathrm{FF}}}{\displaystyle\sum_{(j,v)\neq(k,u)}^{(K,F)} a_{j,v,n} p_{j,v,n}^{\mathrm{F}} g_{k,j,v,n}^{\mathrm{FF}} + p_{w,n}^{\mathrm{M}} g_{k,w,n}^{\mathrm{FM}} + \sigma^2} \tag{3.1}$$

where $\displaystyle\sum_{(j,v)\neq(k,u)}^{(K,F)} a_{j,v,n} p_{j,v,n}^{\mathrm{F}} g_{k,j,v,n}^{\mathrm{FF}} = \sum_{j=1}^{K} \sum_{v=1}^{F} a_{j,v,n} p_{j,v,n}^{\mathrm{F}} g_{k,j,v,n}^{\mathrm{FF}} - a_{k,u,n} p_{k,u,n}^{\mathrm{F}} g_{k,k,u,n}^{\mathrm{FF}}$ is the interference caused by other co-channel femtocells, $p_{w,n}^{\mathrm{M}} g_{k,w,n}^{\mathrm{FM}}$ is the interference caused by the macrocell, and σ^2 is the AWGN power.

The SINR for macro user w using the nth subchannel is given by

$$\gamma_{w,n}^{\mathrm{M}} = \frac{p_{w,n}^{\mathrm{M}} g_{w,n}^{\mathrm{M}}}{\displaystyle\sum_{j=1}^{K} \sum_{v=1}^{F} a_{j,v,n} p_{j,v,n}^{\mathrm{F}} g_{j,v,n}^{\mathrm{MF}} + \sigma^2} \tag{3.2}$$

Based on Shannon's capacity formula, the capacities on subchannel n of femto user u in femtocell k and macro user w are given respectively by

$$C_{k,u,n}^{\mathrm{F}} = \frac{B}{N} \log_2(1 + \gamma_{k,u,n}^{\mathrm{F}}) \tag{3.3}$$

$$C_{w,n}^{\mathrm{M}} = \frac{B}{N} \log_2(1 + \gamma_{w,n}^{\mathrm{M}}) \tag{3.4}$$

3.2.2 Problem Formulation

The maximization of the total capacity of the K femtocells is formulated as follows.

$$\max \sum_{k=1}^{K} \sum_{u=1}^{F} \sum_{n=1}^{N} C_{k,u,n}^{F} \tag{3.5}$$

$$\text{s.t.} \begin{cases} \sum_{n=1}^{N} a_{k,u,n} p_{k,u,n}^{F} \leq p^{\max}, \forall k, u \\ p_{k,u,n}^{F} \geq 0, \forall k, u, n \\ a_{k,u,n} \in \{0,1\}, \forall k, u, n \\ \sum_{u=1}^{F} a_{k,u,n} \in \{0,1\}, \forall k, n \end{cases} \tag{3.6}$$

where a femto user's total transmit power is constrained by $p^{\max}$, the power allocated to each subchannel is nonnegative, and each subchannel is assigned to no more than one user per femtocell.

It is assumed that the user assignment and power allocation can be performed independently for each subchannel, then the maximization of the total capacity of K femtocells is equivalent to the maximization of the total capacity of the K femtocells on one subchannel, and (3.5) and (3.6) can be simplified to

$$\max \sum_{k=1}^{K} \sum_{u=1}^{F} C_{k,u,n}^{F}, \forall n \tag{3.7}$$

$$\text{s.t.} \begin{cases} p_{k,u,n}^{F} \leq p_{n}^{\max}, \forall k, u, n \\ p_{k,u,n}^{F} \geq 0, \forall k, u, n \\ a_{k,u,n} \in \{0,1\}, \forall k, u, n \\ \sum_{u=1}^{F} a_{k,u,n} \in \{0,1\}, \forall k, n \end{cases} \tag{3.8}$$

where $p_{n}^{\max}$ is the transmit power constraint on subchannel n for a femto user. Without loss of generality, we assume that $p_{n}^{\max} = p^{\max}/N$.

3.3 Interference-Aware Resource Allocation

In this section, we first model the uplink power and subchannel allocation problem in femtocells using a non-cooperative game theory framework [21, 22], where a pricing scheme is imposed on femto users to mitigate the uplink interference caused by femto users to the macrocell. Then based on the non-cooperative game framework, we propose a semi-distributed algorithm for femtocells to assign subchannels to femto users assuming an arbitrary power allocation, and then optimize the power allocation on subchannels based on the obtained subchannel assignment.

3.3.1   A Game Theoretic Framework

Based on the microeconomic theory [23], we model the femtocell uplink resource allocation problem as a femtocell non-cooperative resource allocation game (FNRAG). The K femtocells are considered as selfish, rational players. Each of them tries to maximize its utility without considering the impact on other players. The FNRAG for subchannel n can be expressed as

$$G_n = \langle K, \{A_n, P_n\}, \mu_n^c \rangle, \forall n \tag{3.9}$$

where $K = \{1, \ldots, k, \ldots, K\}, \forall n \in \{1, 2, \ldots, N\}$ is the set of femtocells playing the game; $\{A_n, P_n\}$ is the strategy space of the players, with A_n and P_n being the subchannel assignment space and the power allocation strategy space, respectively; and $\mu_n^c = \{\mu_{1,n}^c, \mu_{2,n}^c, \ldots, \mu_{K,n}^c\}$ is the set of net utility functions of the K players, in which

$$\mu_{k,n}^c = \sum_{u=1}^{F} (C_{k,u,n}^{\mathrm{F}} - \alpha a_{k,u,n} g_{k,u,n}^{\mathrm{MF}} p_{k,u,n}^{\mathrm{F}}) \tag{3.10}$$

where $\alpha (\in \mathbb{R}^+)$ (bps/Watt) is the pricing factor, and the price charged on a femto user is proportional to the uplink interference that it causes to the macrocell. If without the pricing part in the utility function, then a player will tend to maximize its utility by using the maximum transmit power, because $C_{k,u,n}^{\mathrm{F}}$ monotonically increases with $p_{k,u,n}^{\mathrm{F}}$ according to (3.1) and (3.3). This will lead to severe uplink interference to the macrocell.

Given the power and subchannel allocation in all other co-channel femtocells, the net utility function of femtocell k can be rewritten as

$$\mu_{k,n}^c(p_{k,n}, a_{k,n} | P_{-k,n}, A_{-k,n}) = \sum_{u=1}^{F} \left[\frac{B}{N} \log_2 \left(1 + \frac{a_{k,u,n} p_{k,u,n}^{\mathrm{F}} g_{k,k,u,n}^{\mathrm{FF}}}{I_{k,u,n}} \right) - \alpha a_{k,u,n} g_{k,u,n}^{\mathrm{MF}} p_{k,u,n}^{\mathrm{F}} \right] \tag{3.11}$$

where $p_{k,n} = \{p_{k,1,n}^{\mathrm{F}}, p_{k,2,n}^{\mathrm{F}}, \ldots, p_{k,F,n}^{\mathrm{F}}\}$, $a_{k,n} = \{a_{k,1,n}, a_{k,2,n}, \ldots, a_{k,F,n}\}$, $P_{-k,n}$ is the $(K-1) \times F$ matrix $(K-1) \times F$ matrix obtained by removing the kth row from P_n, $A_{-k,n}$ is the $(K-1) \times F$ matrix obtained by removing the kth row from A_n, and

$$I_{k,u,n} = \sum_{(j,v) \neq (k,u)}^{(K,F)} a_{j,v,n} p_{j,v,n}^{\mathrm{F}} g_{k,j,v,n}^{\mathrm{FF}} + p_{w,n}^{\mathrm{M}} g_{k,w,n}^{\mathrm{FM}} + \sigma^2.$$

Definition 1. Given the uplink power allocation and subchannel assignment of all other co-channel femtocells, the best response of femtocell k is given by

$$(\hat{p}_{k,n}, \hat{a}_{k,n}) = \arg \max_{p_{k,n}, a_{k,n}} \mu_{k,n}^c(p_{k,n}, a_{k,n} | P_{-k,n}, A_{-k,n}), \forall k \tag{3.12}$$

In the following subsections, in order to solve the non-cooperative femtocell uplink resource allocation game in a semi-distributed manner, we devise a semi-distributed algorithm for each femtocell to first assign subchannels to its femto users for given power and subchannel allocation in all other femtocells and assuming an arbitrary power allocation of its own, and then optimize the power allocation based on the obtained subchannel assignment.

3.3.2 Interference-Aware Subchannel Allocation

In this subsection, we improve the subchannel allocation method in [24, 25] by pricing femto users according to their interference to macrocell in femtocell subchannel allocation.

Assuming that the kth row of the matrix P_n contains an arbitrary power allocation of femtocell k on subchannel n, and given the power allocation and subchannel assignment of all other femtocells on subchannel n, then the best assignment of subchannel n in femtocell k is given by

$$\hat{a}_{k,n} = \arg\max_{a_{k,n}} \mu_{k,n}^c (a_{k,n}|P_n, A_{-k,n}), \forall k \tag{3.13}$$

According to (3.8), at most one element of $a_{k,n}$ can take value of 1. Therefore, based on (3.11), the problem in (3.13) is equivalent to

$$\hat{u}_{k,n} = \arg\max_u \left[\frac{B}{N}\log_2(1 + \frac{p_{k,u,n}^F g_{k,k,u,n}^{FF}}{I_{k,u,n}}) - \alpha g_{k,u,n}^{MF} p_{k,u,n}^F \right], \forall k,n \tag{3.14}$$

where $\hat{u}_{k,n}$ is the best user for channel n in femtocell k to assign, and the assignment of subchannel n in femtocell k is indicated by $\hat{a}_{k,n} = \{\hat{a}_{k,1,n}, \hat{a}_{k,2,n}, \dots, \hat{a}_{k,F,n}\}$, where

$$\hat{a}_{k,u,n} = \begin{cases} 1, & \text{if } u = \hat{u}_{k,n}, \\ 0, & \text{otherwise.} \end{cases} \tag{3.15}$$

In order to remove the dependence of the subchannel assignment on the assumed arbitrary power allocation, we let $\hat{\gamma}_{k,n} = \max_u \frac{p_{k,u,n}^F g_{k,k,u,n}^{FF}}{I_{k,u,n}}$, and then the transmit power of femto user u corresponding to $\hat{\gamma}_{k,n}$ is given by $\frac{\hat{\gamma}_{k,n} I_{k,u,n}}{g_{k,k,u,n}^{FF}}$. Accordingly, (3.14) can be rewritten as

$$\hat{u}_{k,n} = \arg\max_u \left[\frac{B}{N}\log_2(1 + \hat{\gamma}_{k,n}) - \alpha g_{k,u,n}^{MF} \frac{\hat{\gamma}_{k,n} I_{k,u,n}}{g_{k,k,u,n}^{FF}} \right] = \arg\min_u \frac{g_{k,u,n}^{MF}}{g_{k,k,u,n}^{FF}} I_{k,u,n}, \forall k,n \tag{3.16}$$

3.3.3 Interference-Aware Power Allocation

Once the uplink subchannel assignment has been determined by using (3.15) and (3.16) in each femtocell, the FNRAG in (3.9) can be reduced to a femtocell non-cooperative power allocation game (FNPAG): $G_n' = \langle K, P_n, \mu_n^c \rangle, \forall n$. Since subchannels have been assigned to specific femto users in each cell, we will drop the subscript u for simplicity hereafter.

Definition 2. Denote $\hat{p}_n = \{\hat{p}_{1,n}^F, \hat{p}_{2,n}^F, \ldots, \hat{p}_{K,n}^F\}$ as the optimal transmit power vector of the K co-channel femto users allocated to subchannel n under Nash Equilibrium in the FNPAG G_n', if

$$\mu_{k,n}^c(\hat{p}_{k,n}^F | \hat{p}_{-k,n}, A_n) \geq \mu_{k,n}^c(p_{k,n}^F | \hat{p}_{-k,n}, A_n), \forall p_{k,n}^F \geq 0 \qquad (3.17)$$

where $\hat{p}_{-k,n} = \{\hat{p}_{1,n}^F, \ldots, \hat{p}_{k-1,n}^F, \hat{p}_{k+1,n}^F, \ldots, \hat{p}_{K,n}^F\}$ is the optimal transmit power vector of the $K-1$ co-channel femto users using subchannel n under Nash Equilibrium except for the co-channel femto user in femtocell k, and Nash Equilibrium is defined as the fixed points where no player can improve its utility by changing its strategy unilaterally [23].

Theorem 1. A Nash Equilibrium exists in the FNPAG: $G_n' = \langle K, P_n, \mu_n^c \rangle, \forall n$.

Proof. According to [23], a Nash Equilibrium exists in G_n' if the following two conditions are satisfied:

1. P_n is non-empty, convex and compact in the finite Euclidean space $\mathfrak{R}^{K \times F}$.
2. μ_n^c is continuous and concave with respect to P_n.

Since the power allocated on each subchannel is constrained between zero and the maximum power $p^{\max}$, the power allocation matrix P_n is convex and compact, and condition 1 is satisfied.

For condition 2, it can be seen from (3.11) that μ_n^c is continuous with respect to P_n. To prove the quasi-concave property of (3.11), we take the derivative of (3.11) with respect to $p_{k,n}^F$, and get

$$\frac{\partial \mu_{k,n}^c}{\partial p_{k,n}^F} = \frac{B}{N \ln 2} \frac{g_{k,k,n}^{FF}}{\left(I_{k,n} + p_{k,n}^F g_{k,k,n}^{FF}\right)} - \alpha g_{k,n}^{MF} \qquad (3.18)$$

Taking the second-order derivative of (3.11) with respect to $p_{k,n}^F$ yields

$$\frac{\partial^2 \mu_{k,n}^c}{\partial^2 p_{k,n}^F} = -\frac{B}{N \ln 2} \frac{(g_{k,k,n}^{FF})^2}{\left(I_{k,n} + p_{k,n}^F g_{k,k,n}^{FF}\right)^2} \leq 0 \qquad (3.19)$$

Therefore, $\mu_{k,n}^c$ is a quasi-concave function of $p_{k,n}^F$. Since both conditions 1 and 2 hold, a Nash Equilibrium exists in the FNPAG. This completes the proof. ∎

Lemma 1. The best response of femtocell k to the FNPAG is given by

$$\hat{p}_{k,n} = \hat{p}_{k,n}^F \hat{a}_{k,n} \tag{3.20}$$

$$\hat{p}_{k,n}^F = \left[\frac{B}{N \ln 2} \cdot \frac{1}{\alpha g_{k,n}^{MF}} - \frac{I_{k,n}}{g_{k,k,n}^{FF}} \right]_0^{p_n^{max}} \tag{3.21}$$

where $[x]_a^b = \min\{\max\{a,x\},b\}$.

Proof. The $\hat{p}_{k,n}^F$ in (3.21) is obtained by setting (3.18) to zero and solving the resulting equation for $p_{k,n}^F$. ∎

Since (3.21) should be nonnegative, and the interference price factor α is nonnegative too, we get

$$0 \le \alpha \le \frac{B}{N \ln 2} \cdot \frac{g_{k,k,n}^{FF}}{g_{k,n}^{MF} I_{k,n}} \tag{3.22}$$

Theorem 2. The FNPAG has a unique Nash Equilibrium.

Proof. It can be proved following similar proof in [6,27]. ∎

3.3.4 Semi-distributed Implementation

Since only local information, such as uplink interference and channel gains seen by femto users, is needed in calculating (3.16) and (3.21), the interference-aware femtocell uplink subchannel allocation scheme and power allocation scheme proposed in Sects. 3.3.2 and 3.3.3 can be implemented in a distributed and semi-distributed manner, respectively, as outlined in Algorithm 1.

Note that, $g_{k,j,v,n}^{FF}$ and $g_{k,w,n}^{FM}$ for the uplink can be estimated at femto user u in femtocell k by measuring the downlink channel gain of subchannel n from femtocell j and the macrocell, respectively, and utilizing the symmetry between uplink and downlink channels, or by using the site specific knowledge [6]. Furthermore, it can be assumed that there is a direct wire connection between an FBS and the MBS for the FBS to coordinate with the central MBS [5,7], according to a candidate scheme proposed for 3GPP HeNB mobility enhancement [28].

Algorithm 1 can be implemented by each FBS, which only utilizes local information and has limited interaction with the MBS, therefore, Algorithm 1 is semi-distributed and the practicability is guaranteed.

Algorithm 3 Semi-distributed algorithm to solve FNRAG

1: FBS set: $\mathcal{K} = \{1, 2, \ldots K\}$; Femto user set per femtocell: $\mathcal{F} = \{1, 2, \ldots F\}$.
2: **Interference-Aware Subchannel Allocation**
3: Allocate the same power to each subchannel;
4: Femto user u in femtocell k measures $g_{k,u,n}^{\mathrm{MF}}$, $g_{k,k,u,n}^{\mathrm{FF}}$ and $I_{k,u,n}$, $\forall k, u, n$;
5: $a_{k,u,n} = 0$, $\forall k, u, n$;
6: **for** each FBS **do**
7:     Subchannel set: $\mathcal{N} = \{1, 2, \ldots, N\}$
8:     **for** $u = 1$ to F **do**
9:         a) find $n^* = \arg\min_{n \in \mathcal{N}} \frac{g_{k,u,n}^{\mathrm{MF}}}{g_{k,k,u,n}^{\mathrm{FF}}} I_{k,u,n}$;
10:         b) $a_{k,u,n^*} = 1$;
11:         c) $\mathcal{N} = \mathcal{N} - \{n^*\}$;
12:     **end for**
13:     **while** $\mathcal{N} \neq \phi$ **do**
14:         a) find $(u^*, n^*) = \arg\min_{u \in \mathcal{F}, n \in \mathcal{N}} \frac{g_{k,u,n}^{\mathrm{MF}}}{g_{k,k,u,n}^{\mathrm{FF}}} I_{k,u,n}$;
15:         b) $a_{k,u^*,n^*} = 1$;
16:         c) $\mathcal{N} = \mathcal{N} - \{n^*\}$;
17:     **end while**
18: **end for**
19: **Interference-Aware Power Allocation**
20: **for** each FBS **do**
21:     **for** $n = 1$ to N **do**
22:         calculate (3.20) and (3.21);
23:     **end for**
24: **end for**

3.4 Simulation Results and Discussion

In this section, we present simulation results to evaluate the performance of the proposed interference-aware FNRAG algorithm, as compared with the unpriced suboptimal subchannel allocation (USSA) and MIWF-based power allocation algorithm [5, 29], which are outlined in Algorithm 2. Both the system capacity and the fairness between femto-tier and macro-tier are evaluated in the simulations.

In the simulations, the macrocell has a coverage radius of 500 m. Each femtocell has a coverage radius of 10 m. K FBSs and 50 macro users are randomly distributed in the macrocell coverage area. The minimum distance between the MBS and a macro user (or an FBS) is 50 m. The minimum distance between FBSs is 40 m. Femto users are uniformly distributed in the coverage area of their serving femtocell. Both macro and femto cells employ a carrier frequency of 2 GHz, $B = 10$ MHz, and $N = 50$. The AWGN variance is given by $\sigma^2 = \frac{B}{N} N_0$, where

---

Algorithm 4 USSA and MIWF algorithm

1: FBS set: $\mathcal{K} = \{1, 2, \ldots K\}$; Femto user set per femtocell: $\mathscr{F} = \{1, 2, \ldots F\}$.
2: **USSA**
3: Allocate the same power to each subchannel;
4: Femto user u in femtocell k measures $g_{k,k,u,n}^{\mathrm{FF}}$ and $I_{k,u,n}$, $\forall k, u, n$;
5: $a_{k,u,n} = 0$, $\forall k, u, n$;
6: **for** $k = 1$ to K **do**
7:     Subchannel set: $\mathscr{N} = \{1, 2, \ldots, N\}$
8:     **for** $u = 1$ to F **do**
9:         a) find $n^* = \arg \max\limits_{n \in \mathscr{N}} \frac{g_{k,k,u,n}^{\mathrm{FF}}}{I_{k,u,n}}$;
10:         b) $a_{k,u,n^*} = 1$;
11:         c) $\mathscr{N} = \mathscr{N} - \{n^*\}$;
12:     **end for**
13:     **while** $\mathscr{N} \neq \phi$ **do**
14:         a) find $(u^*, n^*) = \arg \max\limits_{u \in \mathscr{F}, n \in \mathscr{N}} \frac{g_{k,k,u,n}^{\mathrm{FF}}}{I_{k,u,n}}$;
15:         b) $a_{k,u^*,n^*} = 1$;
16:         c) $\mathscr{N} = \mathscr{N} - \{n^*\}$;
17:     **end while**
18: **end for**
19: **MIWF-Based Power Allocation**
20: Implement the MIWF algorithm using bisection search [29].

$N_0 = -174 \, \mathrm{dBm/Hz}$. The Rayleigh-fading channel gains are modeled as unit-mean exponentially distributed random variables. The average channel gain (including pathloss and antenna gains) for indoor femto user and outdoor macro user are modeled as λd^{-4} and λd^{-3}, respectively, where $\lambda = 2 \times 10^{-4}$[6]. Besides, α is selected as 4×10^4 using the try-and-error method through simulations. The maximum transmit powers of a femto user and a macro user are set as 20 and 30 dBm, respectively.

Figure 3.2 shows the capacity of the macrocell when the number of femto users per femtocell increases from 1 to 6, for $K = 20, 30$, and 50. It can be observed that the proposed interference-aware FNRAG algorithm outperforms the USSA and MIWF based algorithm by up to a 23% increase in macrocell capacity. As the number of femtocells K increases, the advantage of the FNRAG algorithm becomes more noticeable, because the increased uplink interference caused by femtocells to the macrocell can be effectively mitigated by the pricing scheme imposed on femto users in the FNRAG algorithm, but not by the unpriced USSA and MIWF based algorithm. As the number of femto users increases, the potential interferers will be more because of the co-channel deployed femtocells, but the number of available channels is constant. Therefore, the proposed algorithm will be more and more superior compared with the USSA and MIWF based algorithm. It also can be seen from the figure, the capacity of the macrocells increases as the number of the femto users per femtocell increases. This is because, all of the N subchannels will be assigned to F femto users in each femtocell, when F increases, Algorithm 1 will have more choice for each subchannel in subchannel allocation part, which can be seen as multiuser diversity.

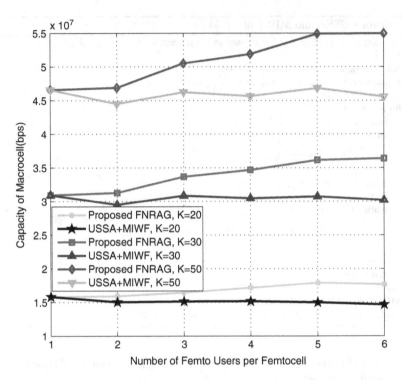

Fig. 3.2 Capacity of macrocell versus the number of femto users per femtocell F with the number of macro users $M = 50$

Figure 3.3 shows the total capacity of K femtocells and macrocell when the number of femto users per femtocell increases from 1 to 6, for $K = 20$, 30, and 50. We can see that the proposed FNRAG algorithm improves the total capacity of femtocells and macrocell over the USSA and MIWF based algorithm by $5 \sim 10\%$ when the number of femto users per femtocell is larger than 3. More gain is obtained as K increases, indicating that the proposed interference-aware FNRAG algorithm can also effectively mitigate interference between neighboring femtocells, and hence is more applicable in dense deployment of co-channel femtocells than the USSA and MIWF based algorithm. As the number of femto users increases, the co-tier interference between femtocells is more severe, the interference-aware subchannel assignment will be more effective in co-tier interference mitigation.

In order to evaluate the fairness between the macro tier and femto tier, we use the tiered fairness index (TFI) [30], which is defined as

$$f_{TFI} = \frac{\left(M \sum_{w=1}^{M} C_w^{\mathrm{M}} + F \sum_{k=1}^{K} \sum_{u=1}^{F} C_{k,u}^{\mathrm{F}} \right)^2}{(M + FK) \left[\sum_{w=1}^{M} \left(M C_w^{\mathrm{M}} \right)^2 + \sum_{k=1}^{K} \sum_{u=1}^{F} \left(F C_{k,u}^{\mathrm{F}} \right)^2 \right]} \tag{3.23}$$

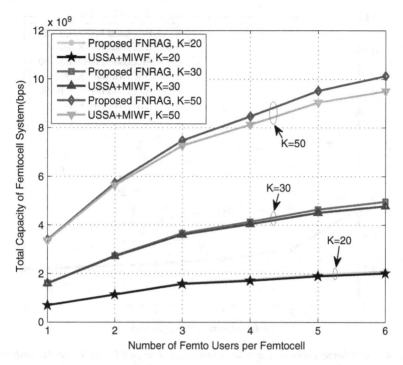

Fig. 3.3 Capacity of macrocell and femtocells versus the number of femto users per femtocell F with the number of macro users $M = 50$

where C_w^M and $C_{k,u}^F$ are the capacities of macro user w and femto user u in femtocell k, respectively.

Figure 3.4 compares the tiered fairness performance between the proposed FNRAG algorithm and the USSA and MIWF based algorithm. It can be observed that the tiered fairness of the proposed FNRAG algorithm gets close to or becomes even better than that of the USSA and MIWF based algorithm, as the number of femto users per femtocell goes beyond 3. This is because the proposed FNRAG algorithm alleviates the uplink interference generated by femto users to the macrocell by charging each femto user a price proportional to the interference that it causes to the macrocell, and the macrocell capacity and consequently the tiered fairness can be guaranteed. Since the USSA used in Algorithm 2 considers the fairness among femto users in each femtocell, the tiered fairness of Algorithm 2 is better than the proposed FNRAG algorithm when F is less than 3. The tiered fairness improves as K increases, because the spatial reuse gain increases with K.

3.5 Conclusion

In this chapter, we have proposed a semi-distributed interference-aware resource allocation algorithm for the uplink of co-channel deployed femtocells, based on a

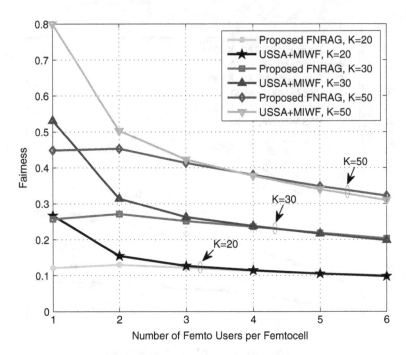

Fig. 3.4 Tiered fairness versus the number of femto users F in each femtocell with the number of macro users $M = 50$

non-cooperative game framework. Using the proposed algorithm, each femtocell can maximize its capacity through resource allocation, taking into account inter-cell interference reported by its femto users, and with uplink femto-to-macro interference alleviated by a pricing scheme imposed on femto users. It has been shown through simulations that the proposed interference-aware resource allocation algorithm is able to provide improved capacities of both macrocell and femtocells, together with comparable tiered fairness, as compared with the existing unpriced subchannel allocation and MIWF based power allocation algorithm.

Acknowledgements The authors would like to thank Dr. David López-Pérez and Prof. Arumugam Nallanathan for their helpful discussions. This work was supported by the Sci-tech Projects of the Committee on Science and Technology of Beijing (D08080100620802, Z101101004310002), the National Natural Science Foundation of China (61101109), and National Key Technology R&D Program of China (2010ZX03003-001-01, 2011ZX03003-002-01), Co-building Project of Beijing Municipal Education Commission "G-RAN based Experimental Platform for Future Mobile Communications", "Research on Resource Allocation and Scheduling Strategy of Future Wireless Communication System" and "Cooperative Communications Platform for Multi-agent Multimedia Communications", Key Fund of Beijing Key Laboratory on Future Network Research. This work was also partially supported by the UK EPSRC Grants EP/H020268/1, CASE/CNA/07/106, and EP/G042713/1.

References

1. D. Lopez-Perez, A. Valcarce, G. de la Roche, and J. Zhang, "Ofdma femtocells: A roadmap on interference avoidance," *IEEE Communications Magazine*, vol. 47, no. 9, pp. 41–48, 2009.
2. *E-UTRA and E-UTRAN Overall Description*, 3GPP Std. TS 36.300 v10.0.0, 2010.
3. V. Chandrasekhar and J. G. Andrews, "Femtocell networks: A survey," *IEEE Commun. Mag.*, vol. 46, no. 9, pp. 59–67, 2008.
4. V. Chandrasekhar, J. Andrews, andA. Gatherer, "Femtocell networks: a survey," *Communications Magazine, IEEE*, vol. 46, no. 9, pp. 59–67, september 2008.
5. K. Son, S. Lee, Y. Yi, and S. Chong, "Refim: A practical interference management in heterogeneous wireless access networks," *IEEE Journal on Selected Areas in Communications*, vol. 29, no. 6, pp. 1260–1272, 2011.
6. V. Chandrasekhar, J. G. Andrews, T. Muharemovic, Z. Shen, and A. Gatherer, "Power control in two-tier femtocell networks," *IEEE Transactions on Wireless Communications*, vol. 8, no. 8, pp. 4316–4328, 2009.
7. X. Kang, R. Zhang, and M. Motani, "Price-based resource allocation for spectrum-sharing femtocell networks: a stackelberg game approach," *IEEE J. Sel. Areas in Commun.*, 2012.
8. H.-S. Jo, C. Mun, J. Moon, and J.-G. Yook, "Interference mitigation using uplink power control for two-tier femtocell networks," *Wireless Communications, IEEE Transactions on*, vol. 8, no. 10, pp. 4906–4910, october 2009.
9. E. J. Hong, S. Y. Yun, and D.-H. Cho, "Decentralized power control scheme in femtocell networks : A game theoretic approach," in *IEEE PIMRC'09*, pp. 1–5.
10. I. Guvenc, M.-R. Jeong, F. Watanabe, and H. Inamura, "A hybrid frequency assignment for femtocells and coverage area analysis for co-channel operation," *Communications Letters, IEEE*, vol. 12, no. 12, pp. 880 –882, december 2008.
11. C. Lee, J.-H. Huang, and L.-C. Wang, "Distributed channel selection principles for femtocells with two-tier interference," in *Vehicular Technology Conference (VTC 2010-Spring), 2010 IEEE 71st*, may 2010, pp. 1–5.
12. I. Mustika, K. Yamamoto, H. Murata, and S. Yoshida, "Potential game approach for self-organized interference management in closed access femtocell networks," in *Vehicular Technology Conference (VTC Spring), 2011 IEEE 73rd*, may 2011, pp. 1–5.
13. J. Kim and D.-H. Cho, "A joint power and subchannel allocation scheme maximizing system capacity in indoor dense mobile communication systems," *IEEE Transactions on Vehicular Technology*, vol. 59, no. 9, pp. 4340–4353, 2010.
14. K. Lee, H. Lee, and D.-H. Cho, "Collaborative resource allocation for self-healing in self-organizing networks," in *IEEE ICC'11*, pp. 1–5.
15. J. Zhang, Z. Zhang, K. Wu, and A. Huang, "Optimal distributed subchannel, rate and power allocation algorithm in ofdm-based two-tier femtocell networks," in *IEEE VTC'10*, pp. 1–5.
16. L. Giupponi and C. Ibars, "Distributed interference control in ofdma-based femtocells," in *IEEE PIMRC'10*, pp. 1201–1206.
17. J.-H. Yun and K. G. Shin, "Adaptive interference management of ofdma femtocells for co-channel deployment," *IEEE Journal on Selected Areas in Communications*, vol. 29, no. 6, pp. 1225–1241, 2011.
18. H. Zhang, X. Chu, W. Ma, W. Zheng and X. Wen, "Resource Allocation with Interference Mitigation in OFDMA Femtocells for Co-channel Deployment," EURASIP Journal on Wireless Communications and Networking: Special Issue on Femtocells in 4G Systems, 2012:289, doi:10.1186/1687-1499-2012-289.
19. H. Zhang, X. Chu, W. Zheng and X. Wen, "Interference-Aware Resource Allocation in Two-Tier OFDMA Co-Channel Deployed Femtocell Networks," to appear in Proc. of IEEE ICC'12, Ottawa, Canada, June 2012.
20. S. Yun, Y. Yi, D.-H. Cho, and J. Mo, "Open or close: On the sharing of femtocells," in *IEEE INFOCOM'11*, pp. 116–120.

21. H. Kwon and B. G. Lee, "Distributed resource allocation through noncooperative game approach in multi-cell ofdma systems," in *IEEE ICC'06*, pp. 4345–4350.
22. Z. Liang, Y. H. Chew, and C. C. Ko, "On the modeling of a non-cooperative multicell ofdma resource allocation game with integer bit-loading," in *IEEE Globecom'09*, pp. 1–5.
23. D. Fudenberg and J. Tirole, *Game Theory*. MIT Press, 1993.
24. C. U. Saraydar, N. B. Mandayam, and D. J. Goodman, "pricing and power control in a multicell wireless data network," *IEEE Journal on Selected Areas in Communications*, vol. 19, no. 10, pp. 1883–1892, 2001.
25. F. Wang, M. Krunz, and S. Cui, "Price-based spectrum management in cognitive radio networks," *IEEE Journal of Selected Topics in Signal Processing*, vol. 2, no. 1, pp. 74–87, 2008.
26. C. Zhong, C. Li, and L. Yang, "Dynamic resource allocation algorithm for multi-cell ofdma systems based on noncoomperative game theory," *Journal of Electronics & Information Technology*, vol. 31, no. 8, pp. 1935–1940, 2009.
27. C. Tan, T. Chuah, and S. Tan, "Fair subcarrier and power allocation for multiuser orthogonal frequency-division multiple access cognitive radio networks using a colonel blotto game," *IET Communications*, vol. 5, no. 11, pp. 1607–1618, 2011.
28. *Way forward proposal for (H)eNB to HeNB mobility*, 3GPP Std. R3-101 849, 2010.
29. W. Yu, "Sum-capacity computation for the gaussian vector broadcast channel via dual decomposition," *IEEE Transactions on Information Theory*, vol. 52, no. 2, pp. 754–759, 2006.
30. M. C. Erturk, H. Aki, I. Guvenc, and H. Arslan, "Fair and qos-oriented spectrum splitting in macrocell-femtocell networks," in *IEEE Globecom'10*, pp. 1–6.

Chapter 4
Resource Allocation in Femtocells with Cross-Tier Interference Limits

Abstract In this chapter, we consider the joint subchannel and power allocation problem in both the uplink and the downlink for two-tier networks comprising spectrum-sharing macrocells and femtocells. A joint subchannel and power allocation scheme for co-channel femtocells is proposed, aiming to maximize the capacity for delay-tolerant users subject to delay-sensitive users' quality of service and interference constraints imposed by macrocells. The joint subchannel and power allocation problem is modeled as a mixed integer programming problem, then transformed into a convex optimization problem by relaxing subchannel sharing, and finally solved by a dual decomposition approach. The effectiveness of the proposed approach is verified by simulations and compared with existing scheme.

4.1 Introduction

Femtocells are low power, low cost, user deployed wireless access points that use local broadband connections as backhaul and compensate macrocells' drawbacks on indoor coverage [1]. Orthogonal frequency division multiple access (OFDMA) based femtocells have been already considered in major wireless communication standards, e.g., LTE/LTE-Advanced [1]. Due to the scarcity of spectrum, operators prefer spectrum sharing between macrocells and femtocells rather than orthogonal deployments [2]. However, cross-tier interference could be severe in spectrum sharing two-tier networks [3]. As a result, resource allocation considering cross-tier interference has become an important asset to enhance performance and has attracted much attention within the telecommunication industry.

H. Zhang et al., *4G Femtocells: Resource Allocation and Interference Management*,
SpringerBriefs in Computer Science, DOI 10.1007/978-1-4614-9080-7_4,
© The Author(s) 2013

Interference mitigation based on resource allocation has been widely analyzed to maintain user's quality of service (QoS), e.g., signal to interference and noise ratio (SINR) capacity, while alleviating cross-tier interference in two-tier networks. In [4], a non-cooperative power allocation with SINR adaptation is used to alleviate the uplink interference suffered by macrocells; while in [5], a Stackelberg game based power control is formulated to maximize femtocell's capacity under cross-tier interference constraints. However, subchannel allocation is not considered. In [6], a joint subchannel and power allocation algorithm is proposed to maximize total capacity in dense femtocell deployments. While in [7], a Lagrangian dual decomposition based resource allocation scheme with constraints on cross-tier interference in power allocations is used. In [2], the distributed subchannel and power allocation for co-channel deployed femtocells is modeled as a non-cooperative game, for which a Nash Equilibrium is obtained based on a time-sharing subchannel allocation. However, in these works, joint subchannel and power allocation with users' QoS and cross-tier interference considerations is not studied. In [8], a distributed modulation and coding scheme, subchannel and power allocation that supports different throughput constraints per users is proposed, but it does not consider two-tier networks.

Femtocell networks should support the heterogeneous QoS for the delay sensitive services such as online gaming and video phone calls, while maximizing the throughput of delay tolerant services [9]. However, to the best of our knowledge, resource allocation for heterogeneous QoS users in femtocell has not been studied in previous works. Indeed, interference mitigation via resource allocation strategies have been widely studied in spectrum underlay Cognitive Radio (CR) networks [10, 11], but cannot be directly applied in femtocells [5].

In this chapter, we focus on the subchannel and power allocation problem in OFDMA based two-tier femtocell networks, in which a central macrocell is underlaid with spectrum-sharing deployed femtocells. Heterogeneous QoS requirement for femto users is considered, where delay sensitive users have a minimum data rate requirement and delay tolerant users do not have. After introducing the interference temperature limit, the resource allocation problem is formulated into a mixed integer non-convex programming problem. To transform this non-convex problem into a convex one, time-sharing subchannel scheme is introduced. Next, we solve the joint subchannel and power allocation problem using Lagrange dual decomposition approach, and devise a distributed joint power and subchannel allocation algorithm. Furthermore, a low-complexity approach is proposed to trade the performance for computational complexity, a significant reduction in computational burden is achieved by the proposed algorithm [12]. The complexity of the proposed algorithm is analyzed, and the performance of the approach is verified by simulations.

The rest of this chapter is organized as follows. Section 4.2 provides the system model and the problem formulation of resource allocation. In Sect. 4.3, the subchannel and power allocation algorithm based on dual decomposition method is proposed. In Sect. 4.4, performance of the proposed algorithm is evaluated by simulations. Finally, Sect. 4.5 concludes the chapter.

4.2 System Model and Problem Formulation

4.2.1 System Model

We consider a two-tier OFDMA network where K co-channel Femto Base Stations (FBSs) are overlaid on a macrocell. All femtocells are assumed to be closed access and deployed in suburban residential houses [13]. We focus on the resource allocation in the uplink of femtocells, and then extend it to the downlink case. Let M and F denote the numbers of active macro users camping on the macrocell and femto users camping on each femtocell, respectively. The OFDMA system has a bandwidth of B, which is divided into N subchannels. The channel fading of each subcarrier is assumed the same within a subchannel, but may vary cross different subchannels. Channel fading is composed of large-scale fading (path loss) and small-scale fading (frequency-selective Rayleigh fading).

The received signal-to-interference-plus-noise ratio (SINR) $\gamma_{k,u,n}^F$ at the kth FBS from its femto user u with $(u \in \{1,2,\ldots,F\})$ in the nth subchannel ($n \in \{1,2,\ldots,N\}$) is modeled as:

$$\gamma_{k,u,n}^F = \frac{p_{k,u,n}^F g_{k,u,n}^F}{p_{w,n}^M g_{k,w,n}^{FM} + \sigma^2},$$
(4.1)

where $g_{k,u,n}^F$ is the channel gain on subchannel n from femto user $u (u \in \{1,2,\ldots,F\})$ to its serving femtocell k; $g_{k,w,n}^{FM}$ is the channel gain on subchannel n from macro user $w(w \in \{1,2,\ldots,M\})$ to femtocell k; $p_{k,u,n}^F$ is femto user u's transmit power on subchannel n in femtocell k; $p_{w,n}^M$ is macro user w's transmit power on subchannel n in the macrocell; and σ^2 is the additive white Gaussian noise (AWGN) power. In this case, co-channel interference between femtocells is assumed as part of the thermal noise due to the severe wall penetration loss and low power of FBSs [7, 14]. Especially, in sparse deployment of femtocells in suburban environments [5], co-tier inter-femtocell interference is negligible as compared with cross-tier interference [14, 15].

Based on Shannon's capacity formula, the uplink capacity of femto user u in femtocell k on subchannel n is modeled by:

$$C_{k,u,n}^F = \log_2(1 + \gamma_{k,u,n}^F).$$
(4.2)

4.2.2 Problem Formulation

Our target is to maximize the total capacity of delay-tolerant users in the K femtocells under macrocell's interference constraints and delay-sensitive users' QoS constraints. The corresponding problem for the uplink can be formulated as the following non-convex mixed integer programming problem:

$$\max_{a_{k,u,n}, p^{\mathrm{F}}_{k,u,n}} \sum_{k=1}^{K} \sum_{u \in \mathscr{D}\mathscr{T}_k} \sum_{n=1}^{N} a_{k,u,n} C^{\mathrm{F}}_{k,u,n} \tag{4.3}$$

s.t. $C1 : \displaystyle\sum_{n=1}^{N} a_{k,u,n} p^{\mathrm{F}}_{k,u,n} \leq P_{\max}, \forall k, u$

$C2 : p^{\mathrm{F}}_{k,u,n} \geq 0, \forall k, u, n$

$C3 : \displaystyle\sum_{n=1}^{N} a_{k,u,n} C^{\mathrm{F}}_{k,u,n} \geq R_u, \forall k, \forall u \in DS_k$

$$\tag{4.4}$$

$C4 : \displaystyle\sum_{k=1}^{K} \sum_{u=1}^{F} a_{k,u,n} p^{\mathrm{F}}_{k,u,n} g^{\mathrm{MF}}_{k,u,n} \leq I^{th}_n, \forall n$

$C5 : \displaystyle\sum_{u=1}^{F} a_{k,u,n} \leq 1, \forall k, n$

$C6 : a_{k,u,n} \in \{0, 1\}, \forall k, u, n \,,$

where $\mathscr{D}\mathscr{S}_k$ and $\mathscr{D}\mathscr{T}_k$ are the sets of delay-sensitive and delay-tolerant users camping on FBS k, respectively, where $|\mathscr{D}\mathscr{S}_k| + |\mathscr{D}\mathscr{T}_k| = F$, $\mathscr{D}\mathscr{S}_k \cap \mathscr{D}\mathscr{T}_k = \phi$, and delay-sensitive femto user u has a minimum throughput requirement R_u; $g^{\mathrm{MF}}_{k,u,n}$ is the channel gain on subchannel n from user u in femtocell k to the MBS; $P = [p^{\mathrm{F}}_{k,u,n}]_{K \times F \times N}$ is the power allocation matrix of the K femtocells; and $A = [a_{k,u,n}]_{K \times F \times N}$ is the subchannel indication matrix, being $a_{k,u,n} = 1$ if subchannel n is assigned to femto user u in femtocell k, and $a_{k,u,n} = 0$ otherwise; constraint C1 limits the transmit power of each femto user to be below $P_{\max}$; C2 represents the non-negative power constraint of the transmit power on each subchannel; C3 guarantees the QoS requirement R_u of delay-sensitive femto user u in femtocell k; C4 represents the tolerable interference temperature level on each subchannel of macrocell, with I^{th}_n as the interference threshold; C5 and C6 guarantee that each subchannel can only be assigned to at most one user in each femtocell.

4.3 Subchannel and Power Allocation Algorithm

In this section, in order to reduce complexity, firstly, we transform the non-convex mixed integer programming problem in (4.3)–(4.4) into a convex optimization by relaxing subchannel allocations. Then, we solve the resulting subchannel and power allocation problem using the Lagrangian dual decomposition method [16, 17].

4.3.1 Transformation of Optimization Problem by Time-Sharing Relaxation

The optimization problem in (4.3) is a non-convex mixed integer programming problem due to the integer constraint in C6. The optimal solution of (4.3) under constraints of (4.4) can be obtained by brute-force or using integer linear programming solvers. However, their running times may be unpredictable (exponential in the worse case). To make the problem tractable, we relax $a_{k,u,n}$ to a continuous real variable in the range $[0,1]$, where $a_{k,u,n}$ can be considered as a time-sharing factor for subchannel n, and can be interpreted as the fraction of time that subchannel n is assigned to femto user k during one transmission frame. The time-sharing relaxation was first proposed in [18], which has been proved to result in a zero-duality gap [19] and has been widely used to transform non-convex combinatorial optimization problems into convex optimization problems for OFDMA systems [2, 9]. For notational brevity, let us denote $s_{k,u,n} = a_{k,u,n} p_{k,u,n}^{\text{F}}$ as the actual power allocated to user u in femtocell k on subchannel n. and $I_{k,u,n} = p_{w,n}^{\text{M}} g_{k,w,n}^{\text{FM}} + \sigma^2$ and $\tilde{C}_{k,u,n}^{\text{F}} = \log_2\left(1 + \frac{s_{k,u,n} g_{k,u,n}^{\text{F}}}{a_{k,u,n} I_{k,u,n}}\right)$ as the received interference and capacity of user u on subchannel n in femtocell k, respectively. As a result, the original problem can be converted into:

$$\max_{a_{k,u,n}, s_{k,u,n}^{\text{F}}} \sum_{k=1}^{K} \sum_{u \in DT_k} \sum_{n=1}^{N} a_{k,u,n} \tilde{C}_{k,u,n}^{\text{F}} \tag{4.5}$$

$$\text{s.t.} \quad C1: \sum_{n=1}^{N} s_{k,u,n} \leq P_{\max}, \forall k, u$$

$$C2: s_{k,u,n} \geq 0, \forall k, u, n$$

$$C3: \sum_{n=1}^{N} a_{k,u,n} \tilde{C}_{k,u,n}^{\text{F}} \geq R_u, \forall u \in DS_k, \forall k$$

$$C4: \sum_{k=1}^{K} \sum_{u=1}^{F} s_{k,u,n} g_{k,u,n}^{\text{MF}} \leq I_n^{th}, \forall n \tag{4.6}$$

$$C5: \sum_{u=1}^{F} a_{k,u,n} \leq 1, \forall k, n$$

$$C6: a_{k,u,n} \in [0, 1], \forall k, u, n.$$

Since the Hessian matrix of every element $a_{k,u,n} \tilde{C}_{k,u,n}^{\text{F}}$ in the summations of (4.5) with respect to $s_{k,u,n}$ and $a_{k,u,n}$ is negative semi-definite, the objective function (4.5) is concave [16]. As the inequality constraints in (4.6) are convex, the feasible set of the objective function is convex. Being a convex optimization problem, the transformed optimization problem in (4.5) has a unique optimal solution, which may be obtained in polynomial time.

4.3.2   Dual Decomposition Method

In the following, the subchannel and power allocation optimization in (4.5) is solved by using the Lagrangian dual decomposition method. The Lagrangian function is given by

$$
\begin{aligned}
&L(\{a_{k,u,n}\}, \{s_{k,u,n}\}, \lambda, \nu, \mu, \eta) \\
&= \sum_{k=1}^{K} \sum_{u \in DT_k} \sum_{n=1}^{N} a_{k,u,n} \tilde{C}_{k,u,n}^{F} + \sum_{k=1}^{K} \sum_{u=1}^{F} \lambda_{k,u} \left(P_{\max} - \sum_{n=1}^{N} s_{k,u,n} \right) \\
&\quad + \sum_{k=1}^{K} \sum_{u \in DS_k} \nu_{k,u} \left(\sum_{n=1}^{N} a_{k,u,n} \tilde{C}_{k,u,n}^{F} - R_u \right) + \sum_{n=1}^{N} \mu_n \left(I_n^{th} - \sum_{k=1}^{K} \sum_{u=1}^{F} s_{k,u,n} g_{k,u,n}^{MF} \right) \\
&\quad + \sum_{k=1}^{K} \sum_{n=1}^{N} \eta_{k,n} \left(1 - \sum_{u=1}^{F} a_{k,u,n} \right),
\end{aligned}
\tag{4.7}
$$

where λ, ν, μ and η are the Lagrange multipliers (also called dual variables) vectors for the constraints $C1$, $C3$, $C4$ and $C5$ in (4.6), respectively, and the boundary constraints $C2$ and $C6$ in (4.6) will be absorbed in the Karush-Kuhn-Tucker (KKT) conditions [16], as it will be shown later. As a result, the Lagrangian dual function is defined as:

$$
\begin{aligned}
&g(\lambda, \nu, \mu, \eta) \\
&= \max_{\{a_{k,u,n}\}, \{s_{k,u,n}\}} L(\{a_{k,u,n}\}, \{s_{k,u,n}\}, \lambda, \nu, \mu, \eta)
\end{aligned}
\tag{4.8}
$$

The dual problem can be expressed as:

$$
\min_{\lambda, \nu, \mu, \eta} g(\lambda, \nu, \mu, \eta)
\tag{4.9}
$$

$$
\text{s.t.}\quad \lambda, \nu, \mu, \eta \geq 0
\tag{4.10}
$$

We decompose the Lagrangian dual function in (4.7) into a master problem and $K \times N$ subproblems. The dual problem can be solved iteratively with each FBS solving the corresponding local subproblem via local information in each iteration. Accordingly, the Lagrangian function in (4.7) is rewritten as:

$$L(\{a_{k,u,n}\}, \{s_{k,u,n}\}, \lambda, \nu, \mu, \eta)$$

$$= \sum_{k=1}^{K} \sum_{n=1}^{N} L_{k,n}(\{a_{k,u,n}\}, \{s_{k,u,n}\}, \lambda, \nu, \mu, \eta) + \sum_{k=1}^{K} \sum_{u=1}^{F} \lambda_{k,u} P_{\max}$$

$$- \sum_{k=1}^{K} \sum_{u \in \mathscr{DS}_k} \nu_{k,u} R_u + \sum_{n=1}^{N} \mu_n I_n^{th} + \sum_{k=1}^{K} \sum_{n=1}^{N} \eta_{k,n} \qquad (4.11)$$

where

$$L_{k,n}(\{a_{k,u,n}\}, \{s_{k,u,n}\}, \lambda, \nu, \mu, \eta)$$

$$= \sum_{u \in DT_k} a_{k,u,n} \tilde{C}_{k,u,n}^{F} - \sum_{u=1}^{F} \lambda_{k,u} s_{k,u,n} + \sum_{u \in DS_k} \nu_{k,u} a_{k,u,n} \tilde{C}_{k,u,n}^{F} - \sum_{u=1}^{F} \mu_n s_{k,u,n} g_{k,u,n}^{MF} \quad (4.12)$$

$$- \sum_{u=1}^{F} \eta_{k,n} a_{k,u,n} .$$

According to the KKT conditions, the optimal solutions of the subproblems, denoted by $\{\hat{a}_{k,u,n}\}$ and $\{\hat{s}_{k,u,n}\}$, can be obtained as:

$$\frac{\partial L_{k,n}(\ldots)}{\partial \hat{s}_{k,u,n}} \begin{cases} = 0, \hat{s}_{k,u,n} > 0 \\ < 0, \hat{s}_{k,u,n} = 0 \end{cases} \forall k, n \qquad (4.13)$$

$$\frac{\partial L_{k,n}(\ldots)}{\partial \hat{a}_{k,u,n}} \begin{cases} < 0, \hat{a}_{k,u,n} = 0 \\ = 0, 0 < \hat{a}_{k,u,n} < 1 \ \forall k, n \ . \\ > 0, \hat{a}_{k,u,n} = 1 \end{cases} \qquad (4.14)$$

We can obtain the optimal transmit power allocation to femto user u in femtocell k at subchannel n by setting (4.13) equal to 0:

$$\hat{p}_{k,u,n}^{F} = \frac{\hat{s}_{k,u,n}}{a_{k,u,n}} = \left(\frac{1}{\ln 2} \cdot \frac{\nu_{k,u}}{\left(\lambda_{k,u} + \mu_n g_{k,u,n}^{MF}\right)} - \frac{I_{k,u,n}}{g_{k,u,n}^{F}} \right)^{+},$$

$$\forall u \in \mathscr{DS}_k \qquad (4.15)$$

$$\hat{p}_{k,u,n}^{F} = \frac{\hat{s}_{k,u,n}}{a_{k,u,n}} = \left(\frac{1}{\ln 2} \cdot \frac{1}{\left(\lambda_{k,u} + \mu_n g_{k,u,n}^{MF}\right)} - \frac{I_{k,u,n}}{g_{k,u,n}^{F}} \right)^{+},$$

$$\forall u \in \mathscr{DT}_k , \qquad (4.16)$$

where $(x)^+ = \max(0, x)$.

We can obtain the optimal subchannel allocation as follows. In (4.14), the partial derivative of Lagrangian function can be expressed as:

$$\frac{\partial L_{k,n}(\ldots)}{\partial \hat{a}_{k,u,n}} = H_{k,u,n} - \eta_{k,n} \qquad (4.17)$$

where

$$
\begin{aligned}
H_{k,u,n} = \\
\tilde{v}_{k,u}\left(\log_2\left(1+\frac{\hat{p}^{F}_{k,u,n}g^{F}_{k,u,n}}{I_{k,u,n}}\right)-\frac{1}{\ln 2}\left(\frac{\hat{p}^{F}_{k,u,n}g^{F}_{k,u,n}}{\hat{p}^{F}_{k,u,n}g^{F}_{k,u,n}+I_{k,u,n}}\right)\right) \\
-\lambda_{k,u}p^{F}_{k,u,n}-\mu_n p^{F}_{k,u,n}g^{MF}_{k,u,n} ,
\end{aligned}
\tag{4.18}
$$

being $\tilde{v}_{k,u}=v_{k,u}, \forall u \in \mathscr{DS}_k$ and $\tilde{v}_{k,u}=1, \forall u \in \mathscr{DT}_k$. Then, subchannel n is assigned to the user u with the largest $H_{k,u,n}$ in femtocell k:

$$
\hat{a}_{k^*,u^*,n}=1\big|_{(k^*,u^*)=\max\limits_{k,u} H_{k,u,n}}, \forall n ,
\tag{4.19}
$$

Finally, based on the subgradient method, the dual variables are updated according to the following expressions:

$$
\lambda^{(i+1)}_{k,u}=\left[\lambda^{(i)}_{k,u}-\beta^{(i)}_1\left(P_{\max}-\sum_{n=1}^{N}s_{k,u,n}\right)\right]^{+}, \forall k,u
\tag{4.20}
$$

$$
v^{(i+1)}_{k,u}=\left[v^{(i)}_{k,u}-\beta^{(i)}_2\left(\sum_{n=1}^{N}\tilde{C}^{F}_{k,u,n}-R_u\right)\right]^{+}, \forall k,u \in \mathscr{DS}_k
\tag{4.21}
$$

$$
\mu^{(i+1)}_n=\left[\mu^{(i)}_n-\beta^{(i)}_3\left(I^{th}_n-\sum_{k=1}^{K}\sum_{u=1}^{F}s_{k,u,n}g^{MF}_{k,u,n}\right)\right]^{+}, \forall n
\tag{4.22}
$$

where $\beta^{(i)}_1, \beta^{(i)}_2$ and $\beta^{(i)}_3$ are the step sizes of iteration $i(i \in \{1,2,\ldots,I_{\max}\})$, $I_{\max}$ is the maximum number of iterations, and the step size should satisfy:

$$
\sum_{i=1}^{\infty}\beta^{(i)}_l=\infty, \lim_{i\to\infty}\beta^{(i)}_l=0, \forall l \in \{1,2,3\}.
\tag{4.23}
$$

4.3.3   Iterative Resource Allocation Algorithm

Equations (4.15)–(4.23) give a solution to the joint subchannel and power allocation problem. The following algorithm implements the solution.

Algorithm 5 Iterative resource allocation algorithm

1: Initialize I_{max} and Lagrangian variables vectors λ, v, μ, set $i = 0$
2: Initialize $p_{k,u,n}$ with a uniform power distribution among all subchannels
3: Initialize $a_{k,u,n}$ with subchannel allocation method in [20], $\forall k, u, n$
4: **repeat**
5:    **for** $k = 1$ to K **do**
6:       **for** $n = 1$ to N **do**
7:          **for** $u = 1$ to F **do**
8:             a) Delay-sensitive users update $\hat{p}^F_{k,u,n}$ according to (4.15);
9:             b) Delay-tolerant users update $\hat{p}^F_{k,u,n}$ according to (4.16);
10:            c) Calculate $H_{k,u,n}$ according to (4.18);
11:            d) FBSs update $\hat{a}_{k,u^*,n}$ according to (4.19), and perform allocation;
12:            e) FBSs update λ, v according to (4.20) and (4.21).
13:         **end for**
14:      **end for**
15:   **end for**
16:   MBS update μ according to (4.22), and broadcast those values to all FBSs via backhaul, $i = i + 1$.
17: **until** Convergence or $i = I_{max}$

The proposed algorithm can be implemented in each FBS, using only local information and limited interactions with the MBS. As a result, it is distributed.

Note that $g^{MF}_{k,u,n}$ required in (4.15)–(4.16), (4.18) and (4.22) for the uplink can be estimated at femto user u in femtocell k by measuring the downlink channel gain of subchannel n from the macrocell and assuming the symmetry between uplink and downlink channels, or by using the site specific knowledge [21]. Moreover, it could be assumed that there is a wired connection between an FBS and the MBS for coordination purposes [3, 5], according to a candidate scheme proposed for 3GPP HeNB mobility enhancement [22].

4.3.4 Downlink Case

Although the above proposed subchannel and power allocation scheme was presented for the uplink, the algorithm derived for the uplink can also be applied in the downlink with some modifications. The major modifications include replacing channel gains of the reverse link with those of the forward link, replacing the total power constraint for a femto user in (4.4) and (4.6) with the power budget of an FBS, and replacing (4.20) with the following dual variable update:

$$\lambda_k^{(i+1)} = \left[\lambda_k^{(i)} - \beta_1^{(i)} \left(P_{k,max} - \sum_{u=1}^{F} \sum_{n=1}^{N} s_{k,u,n} \right) \right]^+, \forall k \qquad (4.24)$$

where $P_{k,max}$ is the maximal transmit power of FBS k. Moreover, when the MBS updates cross-tier interference dual variables according to (4.22), cross-tier interference can be obtained from macro users' feedback.

4.4   Simulation Results and Discussions

Simulation results are given in this section to evaluate the performance of the proposed resource allocation algorithms. In the simulations, spectrum-sharing femtocells and macro users are randomly distributed in the macrocell coverage area, and femto users are uniformly distributed within the coverage area of their serving femtocell. The coverage radius of the macrocell is 500 m, while that of a femtocell is 10 m. Macro and femto users' maximum transmit powers are set at 23 dBm. The carrier frequency is 2 GHz, $B = 10$ MHz, $N = 50$, $M = 50$, and $\sigma^2 = \frac{B}{N}N_0$, where $N_0 = -174$ dBm/Hz is the AWGN power spectral density. The path loss models for indoor femto users and outdoor macro users are based on [23], and block-fading channel gains are modeled as i.i.d. unit-mean exponentially distributed random variables. The standard deviation of shadow fading between the MBS and users is 8 dB, while that between an FBS and users is 10 dB. The "Existing Algorithm" included in simulations for comparison is the subchannel allocation scheme in [20] in conjunction with the optimal power allocation scheme proposed in this chapter.

Figure 4.1 shows the convergence of the average capacity per femtocell of the proposed algorithm versus the number of iterations for both the uplink and downlink cases, where $F = 4$, $K = 10$, $Ru = 9$ bps/Hz (for any u), $P_{max} = 23$ dBm, $P_{k,max} = 20$ dBm (for any k) and $I_n^{th} = 7.5 \times 10^{-14}$ w (-101.2 dBm). From the results, it can be seen that the proposed algorithm takes only a few iterations to converge to stable solutions. After 6 iterations it is almost stable, and after 40 iterations it is stable. This result together with former analysis proves that the proposed algorithm converges, is practical and can be applied to two-tier networks.

Figure 4.2 shows the total capacity of K femtocells when the number of femtocells increases from 10 to 50, for the proposed algorithm and the existing algorithm. The simulation parameters are set as $F = 2$, $Ru = 9$ bps/Hz, $P_{max} = 23$ dBm, $P_{k,max} = 20$ dBm (for any k), and 7.5×10^{-14} w (-101.2 dBm). From the results, it can be seen that the proposed algorithm improves the total capacity of femtocells for the uplink over the existing algorithm by $10 \sim 25$ % when the number of femtocells is larger than 20. While in the downlink, a similar performance is observed but with a lower capacity, because the maximal downlink power of FBS is 3 dB lower than the maximal uplink power of a femto user.

Figure 4.3 shows the total capacity of delay sensitive users in all femtocells when the number of femtocells increases from 10 to 50, for the proposed and existing algorithms. The simulation settings of F, Ru, P_{max}, $P_{k,max}$ and I_n^{th} are same as Fig. 4.2. It can be seen that the existing algorithm provides a higher capacity than the proposed algorithm for delay-sensitive users, because the subchannel allocation in the proposed algorithm maximizes the capacity of delay-torrent users while ensuring a certain level capacity for delay-sensitive users. However, the subchannel allocation in the existing algorithm maximizes the total capacity of all the users and maintains users' fairness by allocating at least a subchannel to each user.

Figure 4.4 shows the total uplink capacity of K femtocells employing the proposed algorithm versus the interference temperature limit, which increases from

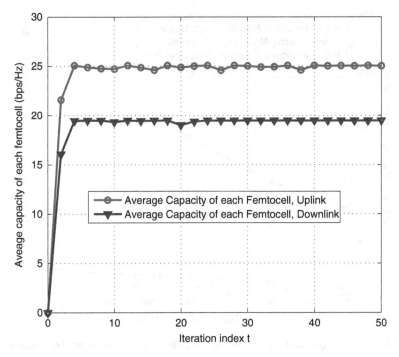

Fig. 4.1 Convergence in terms of average capacity of each femtocell over the number of iterations

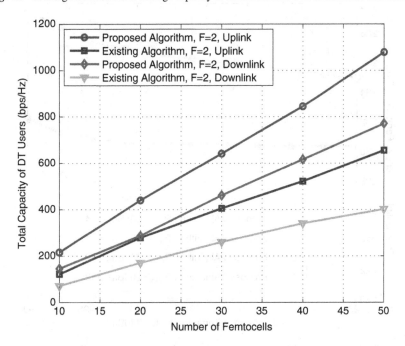

Fig. 4.2 Total capacity of all femtocells versus the number of femtocells K

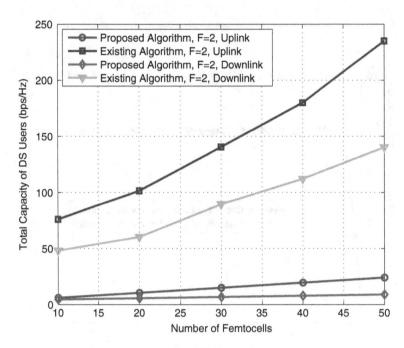

Fig. 4.3 Total capacity of all delay sensitive users in all femtocells versus the number of femtocells

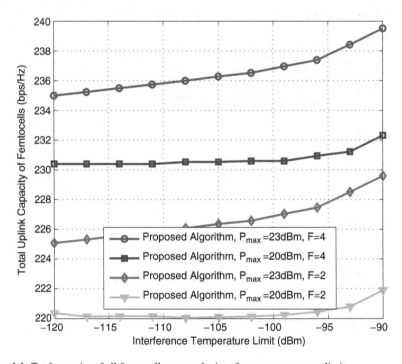

Fig. 4.4 Total capacity of all femtocells versus the interference temperature limit

-120 to -90 dBm, for $P_{max} = 23$ or 20 dBm, $F = 2$ or $F = 4$, $Ru = 9$ bps/Hz, and $K = 10$. It can be seen that as the interference temperature limit increases, the total uplink capacity of femtocells increases, because the tolerable interference caused by femtocells to the macrocell is increased. Moreover, for the proposed algorithm, increasing P_{max} from 20 to 23 dBm increases by up to 5 % the total uplink femtocell capacity, while increasing F from 2 to 4 increases by up to 6 % the femtocell uplink capacity because of multi-user diversity.

4.5 Conclusion

In this chapter, we have investigated the joint subchannel and power allocation for spectrum-sharing femtocells taking heterogeneous femto-user QoS requirements and a cross-tier interference limit into account. The proposed distributed resource allocation algorithm properly allocates resources to users according to their heterogeneous QoS requirements, so as to increase the throughout in the network. Simulation results have shown that the proposed algorithm provides more efficient solutions compared with cutting-edge algorithms in the literature.

Acknowledgements The authors would like to thank Dr. David López-Pérez, Prof. Meixia Tao and Prof. Arumugam Nallanathan for their contributions and helpful discussions. This work was supported by the Co-building Project of Beijing Municipal Education Commission "Research on Resource Allocation and Scheduling Strategy of Future Wireless Communication System" and "Multi-Agent Multimedia Cooperative Communications Platform", Key Fund of Beijing Key Laboratory on Future Network Research, the National Natural Science Foundation of China (61101109), and National Key Technology R&D Program of China (2010ZX03003-001-01, 2011ZX03003-002-01).

References

1. D. López-Pérez, A. Valcarce, G. de la Roche, and J. Zhang, "Ofdma femtocells: A roadmap on interference avoidance," *IEEE Commun. Mag.*, vol. 47, no. 9, pp. 41–48, 2009.
2. J.-H. Yun and K. G. Shin, "Adaptive interference management of ofdma femtocells for co-channel deployment," *IEEE J. Sel. Areas in Commun.*, vol. 29, no. 6, pp. 1225–1241, 2011.
3. K. Son, S. Lee, Y. Yi, and S. Chong, "Refim: A practical interference management in heterogeneous wireless access networks," *IEEE J. Sel. Areas in Commun.*, vol. 29, no. 6, pp. 1260–1272, 2011.
4. V. Chandrasekhar, J. G. Andrews, T. Muharemovic, Z. Shen, and A. Gatherer, "Power control in two-tier femtocell networks," *IEEE Trans. Wireless Commun.*, vol. 8, no. 8, pp. 4316–4328, 2009.
5. X. Kang, R. Zhang, and M. Motani, "Price-based resource allocation for spectrum-sharing femtocell networks: a stackelberg game approach," *IEEE J. Sel. Areas in Commun.*, 2012.
6. J. Kim and D.-H. Cho, "A joint power and subchannel allocation scheme maximizing system capacity in indoor dense mobile communication systems," *IEEE Trans. Veh. Technol.*, vol. 59, no. 9, pp. 4340–4353, 2010.

7. J. Zhang, Z. Zhang, K. Wu, and A. Huang, "Optimal distributed subchannel, rate and power allocation algorithm in ofdm-based two-tier femtocell networks," in *Proc. Veh. Technol. Conf.*, May 2010, pp. 1–5.
8. D. Lopez-Perez, A. Ladanyi, A. Juttner, H. Rivano, and J. Zhang, "Optimization method for the joint allocation of modulation schemes, coding rates, resource blocks and power in self-organizing lte networks," in *INFOCOM, 2011 Proceedings IEEE*, april 2011, pp. 111–115.
9. M. Tao, Y.-C. Liang, and F. Zhang, "Resource allocation for delay differentiated traffic in multiuser ofdm systems," *IEEE Trans. Wireless Commun.*, vol. 7, no. 6, pp. 2190–2201, June 2008.
10. D. T. Ngo and T. Le-Ngoc, "Distributed resource allocation for cognitive radio networks with spectrum-sharing constraints," *IEEE Trans. Veh. Technol.*, vol. 60, no. 7, pp. 3436–3449, Sept. 2011.
11. R. Xie, F. R. Yu, and H. Ji, "Dynamic resource allocation for heterogeneous services in cognitive radio networks with imperfect channel sensing," *IEEE Trans. Veh. Technol.*, no. 99, p. 1, 2011.
12. H. Zhang, W. Zheng, X. Chu, X. Wen, M. Tao, A. Nallanathan and D. López-Pérez, "Joint Subchannel and Power Allocation in Interference-Limited OFDMA Femtocells with Heterogeneous QoS Guarantee", to appear in Proc. of IEEE Globecom 2012.
13. G. de la Roche, A. Valcarce, D. Lopez-Perez, and J. Zhang, "Access control mechanisms for femtocells," *Communications Magazine, IEEE*, vol. 48, no. 1, pp. 33–39, january 2010.
14. V. Chandrasekhar and J. G. Andrews, "Femtocell networks: A survey," *IEEE Commun. Mag.*, vol. 46, no. 9, pp. 59–67, 2008.
15. H.-S. Jo, C. Mun, J. Moon, and J.-G. Yook, "Interference mitigation using uplink power control for two-tier femtocell networks," *IEEE Trans. Wireless Commun.*, vol. 8, no. 10, pp. 4906–4910, Oct. 2009.
16. S. Boyd and L. Vandenberghe, *Convex Optimization*. Cambridge University Press, 2004.
17. D. W. K. Ng and R. Schober, "Resource allocation and scheduling in multi-cell ofdma systems with decode-and-forward relaying," *IEEE Trans. Wireless Commun.*, vol. 10, no. 7, pp. 2246–2258, July 2011.
18. C. Y. Wong, R. Cheng, K. Lataief, and R. Murch, "Multiuser ofdm with adaptive subcarrier, bit, and power allocation," *IEEE J. Sel. Areas in Commun.*, vol. 17, no. 10, pp. 1747–1758, Oct. 1999.
19. W. Yu and R. Lui, "Dual methods for nonconvex spectrum optimization of multicarrier systems," *IEEE Trans. Commun.*, vol. 54, no. 7, pp. 1310–1322, July 2006.
20. Z. Shen, J. G. Andrews, and B. L. Evans, "Adaptive resource allocation in multiuser ofdm systems with proportional rate constraints," *IEEE Trans. Wireless Commun.*, vol. 4, no. 6, pp. 2726–2737, Nov. 2005.
21. J. K. Chen, G. de Veciana, and T. S. Rappaport, "Site-specific knowledge and interference measurement for improving frequency allocations in wireless networks," *IEEE Trans. Veh. Technol.*, vol. 58, no. 5, pp. 2366–2377, June 2009.
22. *Way forward proposal for (H)eNB to HeNB mobility*, 3GPP Std. R3-101 849, 2010.
23. *Further Advancements for E-UTRA, Physical Layer Aspects*, 3GPP Std. TR 36.814 v9.0.0, 2010.

Chapter 5
Energy Efficient Power Control in Femtocells with Interference Pricing

Abstract In two-tier femtocell networks, femtocells cause serious cross-tier interference and consume large amounts of energy for its large-scale deployment. In this chapter, we investigate the power control problem for the co-channel deployed femtocells. We first model the uplink power control problem as a non-cooperative game, where co-channel interference is taken into account in maximizing the energy-aware utility. After introducing a price proportional to the cross-tier interference, Pareto improvement can be obtained. Furthermore, based on the non-cooperative game, we devise a distributed power allocation algorithm together with an optimal price seeking algorithm. Simulation results show that the proposed algorithm can improve users' utilities significantly, compared with existing power control algorithms.

5.1 Introduction

Recently, more and more data services occur in indoor environments [1], where the coverage of macrocells may not be good enough because of the wall penetration losses and long transmission distances from the outdoor macrocell base stations. Thanks to femtocells, the shortcoming of macrocells in providing indoor coverage can be overcome.

In practice, there are still some technical challenges to be further addressed before widespread deployment of femtocells. A two-tier macrocell and femtocell network is usually implemented by sharing frequency rather than splitting frequency between the two tiers [6]. Hence cross-tier interference (CTI) and inter-tier interference (ITI) are the key issues in two-tier macrocell and femtocell networks [2]. The mitigation of cross-tier and inter-cell interference has become an interesting research area. Relevant existing works in the literature will be reviewed in the following.

Game theory has been considered to mitigate interference in two-tier networks with co-channel deployment of femtocells. In [3,6], the minimization of co-tier and cross-tier interference though power control based on game theory is investigated.

H. Zhang et al., *4G Femtocells: Resource Allocation and Interference Management*,
SpringerBriefs in Computer Science, DOI 10.1007/978-1-4614-9080-7_5,
© The Author(s) 2013

In [5], the authors introduce a distributed utility-based signal-to-interference-plus-noise ratio (SINR) adaptation algorithm in order to alleviate cross-tier interference caused by co-channel femtocells to the macrocell. In [7], a decentralized femtocell access strategy based on non-cooperative game is proposed to manage the interference between nearby femtocells and from femtocells to macrocells. The authors in [6] propose a distributed power control algorithm for spectrum-sharing two-tier networks using Stackelberg game, which is very effective in distributed power allocation and macrocell protection while requiring minimal network overhead.

In this chapter, we propose an energy-aware power optimization algorithm for uplink power control in two-tier macrocell and femtocell networks [10], which is based on non-cooperative game with cross-tier interference pricing where each femtocell maximizes its own utility. We employ the super-modularity theory to show the existence of Nash equilibrium. The power optimization algorithm allows a distributed implementation where the cross-tier interference pricing can be broadcast by the base station to all the terminals.

The rest of this chapter is organized as follows. Section 5.2 provides the system model of a two-tier macrocell and femtocell network and the problem formulation. In Sect. 5.3, we discuss the non-cooperative power control game with convex pricing and propose a distributed interference-aware power control algorithm. Performance improvement of the proposed algorithm compared with existing schemes is evaluated by simulation in Sect. 5.4. Finally, Sect. 5.5 concludes the chapter.

5.2   System Model and Problem Formulation

5.2.1   System Model

In this chapter, we consider a two-tier femtocell network in Fig. 5.1, where femtocells are densely deployed [5]. A macrocell base station (MBS) B_0 locates in the center of its coverage area, which is a disc area with a radius of R_m. Within the coverage area of the MBS, femtocell base stations (FBSs) ($\{B_i\}(i = 1 \cdots N)$) are located in a square grid of area D_{grid}^2 sq.km with $\sqrt{N}$ femtocells per dimension, at a distance D_f from the MBS. The coverage radius of each femtocell is R_f. Let $D_{i,j}$ denote the distance between transmitting mobile terminal j and BS B_i ($i = 0 \cdots N$). For simplicity, the channel gains are represented by $g_{i,j}$, and the simplified path loss model in [5] is adopted. It is assumed that femtocells and macrocell use the same frequency spectrum, and there is only one scheduled active user during each signaling slot in each cell. Let $i \in \{0, 1, \ldots, N\}$ denotes the scheduled active user connected to its BS B_i. User i's transmit power is represented by p_i Watts. The variance of additive white Gaussian noise (AWGN) is σ^2. Consequently, the received SINR γ_i of femtocell user i at FBS B_i ($i = 1 \cdots N$) can be expressed as

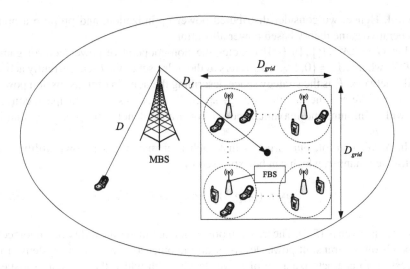

Fig. 5.1 The topology of two-tier femtocell networks

$$\gamma_i = \frac{p_i g_{i,i}}{\sum\limits_{j=1, j\neq i}^{N} p_j g_{i,j} + p_0 g_{i,0} + \sigma^2}, \tag{5.1}$$

where $\sum\limits_{j=1, j\neq i}^{N} p_j g_{i,j}$ is the interference caused by other co-channel femtocells, and $p_0 g_{i,0}$ is the interference caused by the macrocell.

Based on the sigmoid function [8], the UQS function of femtocell user i is defined as

$$f_i(\gamma_i) = \frac{1}{1 + e^{\alpha_i(\beta_i - \gamma_i)}}, \tag{5.2}$$

where α_i and β_i are the UQS parameters [8].

Given interfering powers p_{-i} and the desired transmit power p_i, we model the utility function of user i as

$$u_i(p_i, p_{-i}) = \frac{f_i(\gamma_i)}{p_i} \tag{5.3}$$

where u_i is the energy efficiency of user i.

5.2.2 Problem Formulation

Since a central controller would require complete information of the network, including interference channel gains, it is impractical to use centralized power control. Hence, we consider distributed power optimization, and propose a non-cooperative game theory based power allocation.

Let $G = [\mathcal{N}, \{P_i\}, \{u_i(\cdot)\}]$ denotes the non-cooperative power control game (NPG), where $\mathcal{N} = \{0, 1, \ldots, N\}$ refers to the index set of the users currently active in the network. P_i is the strategy space describing the domain of transmission power for user i. We assume that the strategy space P_i of each user is a compact, convex set with minimum and maximum power constraints denoted by $\underline{p}_i$ and $\bar{p}_i$, that is $P_i = [\underline{p}_i, \bar{p}_i]$.

In power control, our target is that each user maximizes its own utility in a distributed manner, which can be expressed as

$$\max_{p_i \in P_i} u_i(p_i, p_{-i}), \quad \text{for all } i \in \mathcal{N}, \tag{5.4}$$

where u_i is given in (5.3). The transmit power that optimizes the utility of a femtocell depends on the transmit power levels of all the other terminals in the system. It is necessary to characterize a set of power levels, with which the users are satisfied with the utility they receive given the transmit power levels of other users. Such an operating point is called an equilibrium point.

5.3 Non-cooperative Power Control Game with Convex Pricing

The Nash equilibrium of NPG exists and is unique, which has been proved in [10]. In fact, the Nash equilibrium is inefficient in general [10]. In non-cooperative power control game, each terminal selfishly optimizes its own utility, but the interference caused by the terminal will be imposed on other terminals. Thus, the Nash equilibrium is not Pareto efficient. Consequently, we propose a pricing strategy for the power control game by adding a penalty price to each femtocell user's transmission cost. Through pricing, we can depress femtocell users' aggressive behavior and achieve the Pareto improvement by implicitly inducing cooperation, and yet we maintain the non-cooperative nature of the resulting power control solution.

In the existing literature, linear pricing mechanisms with respect to user's uplink transmission power [11–14] have been applied, in order to move the Nash equilibrium solution to a Pareto optimal one. The idea of convex pricing mechanism derives from the observation that the harm a user imposes on other users is not equivalent within the whole range of transmission power, in contrast to the linear pricing mechanism. The main arguments of adopting nonlinear pricing have been summarized in [15].

Based on the nonlinear pricing strategy, we develop a non-cooperative power control game with convex pricing (NPG-CP). We assume that all users participate in a $N + 1$ player non-cooperative power control game $G_c = [\mathcal{N}, \{P_i\}, \{u_i^c(\cdot)\}]$.

5.3.1 Femtocell Utility Function

Each femtocell user seeks to maximize its individual utility, although transmitting with too much power will cause unacceptable cross-tier and co-tier interference. In order to achieve Pareto improvement and reduce the cross-tier interference, given interfering powers p_{-i} and user i's current transmit power p_i, we model the utility function for femtocell user i as

$$u_i^c(p_i, p_{-i}) = \frac{f_i(\gamma_i)}{p_i} - c_i(p_i, p_{-i}) \tag{5.5}$$

where $c_i(p_i, p_{-i})$ is the pricing function for femtocell user i ($i = 1, 2 \ldots, N$).

As described above, the pricing function monotonically increases with the transmit power of the user. Considering cross-tier interference mitigation, we define the convex pricing function as follows

$$c_i(p_i, p_{-i}) = c \cdot e^{g_{0,i} p_i}, \tag{5.6}$$

where c is a positive scalar. Then, (5.5) can be rewritten as

$$u_i^c(p_i, p_{-i}) = \frac{f_i(\gamma_i)}{p_i} - c \cdot e^{g_{0,i} p_i}. \tag{5.7}$$

The pricing discourages femtocell user i from decreasing cellular SINR by transmitting at high power. It is not straightforward to verify whether the Nash equilibrium point exists in NPG-CP, because the femtocell utility functions of NPG-CP are not quasi-concave. Thus, we employ super-modularity theory to prove the existence of Nash equilibrium.

5.3.2 Super-Modular Games

Super-modular game has particular characteristics which can be applied to implement power control schemes. A formal definition of a super-modular game can be found in [16].

Definition 1. A game $G_\varepsilon = [\mathcal{N}, \{P_i\}, \{u_i^\varepsilon(\cdot)\}]$ with parameter ε is said to be super-modular if, $\left(\partial u_i^{\varepsilon 2}(p_i) \big/ \partial p_i \partial p_j\right) \geq 0$ for all $j \neq i$ and $\left(\partial u_i^{\varepsilon 2}(p_i) \big/ \partial p_i \partial \varepsilon\right) \geq 0$ for all i.

The significance of this characteristic is that there exists a fixed point which implies a Nash equilibrium point. However, the NPG-CP $G_c = [\mathcal{N}, \{P_i\}, \{u_i^c(\cdot)\}]$ is not a super-modular game. Next we modify the strategy spaces to transform the game into a super-modular game. Let the compact set $\hat{P}_i = \left[\underline{p}_i, \overline{p}_i\right]$ denote the modified strategy space. Using the condition given in Definition 1, i.e., $\left(\partial u_i^{\varepsilon 2}(p_i)\big/\partial p_i \partial p_j\right) \geq 0$ for all $j \neq i$, we have

$$\frac{\partial u_i(p_i, p_{-i})}{\partial p_i} = \frac{1}{(p_i)^2}\left(\frac{\partial f_i(\gamma_i)}{\partial \gamma_i} \cdot \gamma_i - f_i(\gamma_i)\right) - c g_{0,i} \cdot e^{g_{0,i} p_i}. \tag{5.8}$$

Then,

$$\frac{\partial^2 u_i(p_i, p_{-i})}{\partial p_i \partial p_j} = \frac{1}{(p_i)^2}\frac{\partial^2 f_i(\gamma_i)}{\partial^2 \gamma_i} \cdot \frac{\partial \gamma_i}{\partial p_j} \cdot \gamma_i. \tag{5.9}$$

Obviously, from $\left(\partial u_i^{\varepsilon 2}(p_i)\big/\partial p_i \partial p_j\right) \geq 0$ for all $j \neq i$, we can obtain that $\frac{\partial^2 f_i(\gamma_i)}{\partial^2 \gamma_i} \leq 0$, and

$$\frac{\partial^2 f_i(\gamma_i)}{\partial^2 \gamma_i} = \frac{\alpha_i^2 \cdot e^{\alpha_i(\beta_i - \gamma_i)} \cdot \left(e^{\alpha_i(\beta_i - \gamma_i)} - 1\right)}{\left(1 + e^{\alpha_i(\beta_i - \gamma_i)}\right)^3}. \tag{5.10}$$

Then we can obtain that $\gamma_i \geq \beta_i$, and the smallest power in the modified strategy space $\underline{p}_i$ can be derived from $\gamma_i \geq \beta_i$. We assume that the largest power $\overline{p}_i$ is larger than $\underline{p}_i$.

Theorem 1. The modified game $\hat{G}_c = \left[\mathcal{N}, \{\hat{P}_i\}, \{u_i^c(\cdot)\}\right]$ with parameter c is a super-modular game.

Proof. As discussed above, the condition $(\partial u_i^{\varepsilon 2}(p_i)/\partial p_i \partial p_j) \geq 0$ has been satisfied in the modified strategy space $\hat{P}_i$. Performing a change of variables from c to $-\varepsilon$, we get $(\partial u_i^{\varepsilon 2}(p_i)/\partial p_i \partial \varepsilon) = g_{0,i} \cdot e^{g_{0,i} p_i} \geq 0$ for $i = 1, 2 \dots, N$ and $(\partial u_0^{\varepsilon 2}(p_0)/\partial p_0 \partial \varepsilon) = 0$ for $i = 0$. Therefore, the modified game $\hat{G}_c = [\mathcal{N}, \{\hat{P}_i\}, \{u_i^c(\cdot)\}]$ is a super-modular game.

The set of Nash equilibrium of a super-modular game is nonempty. Furthermore, the Nash-equilibrium set has a largest element and a smallest element. It has been proved in [9]. As discussed above, there exists a Nash equilibrium in the modified game. Let E denote the set of Nash equilibrium and p^S and p^L denote the largest and the smallest elements of E. Note that for two vectors $\mathbf{x}, \mathbf{y} \in \mathcal{R}^n$, $\mathbf{x} > \mathbf{y}$ if and only if $x_i > y_i$ for all $i = 1, 2, \dots n$.

Theorem 2. In the modified NPG-CP, if $p \geq p^S$, where $p, p^S \in E$ and p^S is the smallest elements of E, $u_i^c(p) \leq u_i^c(p^S)$ for all i is p^S the Pareto dominant equilibrium.

Proof. We can observe that, if p_i and c are fixed, utility $u_i^c(p_i, p_{-i})$ decreases with increasing p_{-i} for all i. Since $p_{-i} \geq p_{-i}^S$, we have

$$u_i^c(p_i, p_{-i}) \leq u_i^c(p_i, p_{-i}^S). \tag{5.11}$$

According to the definition of Nash equilibrium and since p^S is a Nash equilibrium of NPG-CP, we have

$$u_i^c \left(p_i, p_{-i}^S \right) \leq u_i^c \left(p_i^s, p_{-i}^S \right). \tag{5.12}$$

From (5.9) and (5.10), we obtain that

$$u_i^c \left(p \right) \leq u_i^c \left(p^S \right). \tag{5.13}$$

It is clear that the best-response of a user terminal is the Nash equilibrium with the minimum total transmission power.

In a super-modular game, if the users' best responses are single-valued, and each user updates starting from the smallest element of its strategy space, then the strategies monotonically converge to the smallest Nash equilibrium, which has been proved in [9]. We denote the smallest Nash equilibrium point as

$$p_i^* = \min\{\min_{p_i} \left(\arg\max_{p_i}(u_i^c) \right), \overline{p}_i\}. \tag{5.14}$$

5.3.3 Energy-Aware Power Control NPG-CP

In this section, we present a distributed power control algorithm that is based on NPG-CP.

Algorithm 6 Iterative distributed power control

1: Initialize the power vector $p(0) = \underline{p}$ at the beginning and the iteration index of power update $k = 1$;
2: For all the iteration times, that is when it satisfies $k \leq K_{\max}$, the power update process will go on according to the following processes:
   i) For all the femtocell user i, given power vector, compute $\gamma_i(k)$ and update its power according to

$$p_i^*(k) = \min\{\min_{p_i} \left(\arg\max_{p_i}(u_i^c) \right), \overline{p}_i\} \qquad i = 1, 2\ldots, N$$

   ii) Assign transmission power according to p_i^*;
3: Iteration index of update increases: $k = k + 1$.
4: If the index exceeds the maximum value $K_{\max}$ or convergence obtains, the power update ends;

The algorithm can be implemented in a distributed manner, that is, each femtocell user i only needs to know its own utility u_i^c, the pricing factor c and its channel gain to B_0 and B_i.

The searching process for the optimal pricing factor value is given as following. First, we implement the Proposed Algorithm (power update) with no pricing $c = 0$. When the equilibrium with no pricing is obtained, we go on implementing power update with pricing Δc, which is a positive value, and then get a set of utilities at equilibrium. Compared with the previous utilities, if the utilities at this new equilibrium increase, then the pricing factor is incremented and the procedure is repeated until the utility is lower than the previous utility for at least one user. The algorithm is summarized in Algorithm 2.

Algorithm 7 Optimal pricing scheme

1: Initialize the pricing factor at the beginning: $c = 0$ that will be known by all femtocell users;
2: Set $\Delta c > 0$ and implement the power update process;
3: Get u_i for all $i \in \mathcal{N}$ at equilibrium, then increase the pricing factor value: $c = c + \Delta c$, which is also known by each user;
4: If $u_i^c \leq u_i^{c+\Delta c}$ for all $i \in \mathcal{N}$ return to step 2, else, that is, if there is one utility worse than the previous utility, the seeking update will stop, and the optimal pricing value is: $c_{optimal} = c$.

5.4 Performance Evaluation

Simulation results are given in this section. A non-pricing power control algorithm (NPG-NP), power optimization with cellular link protection (POCP) [5], and equal power allocation scheme (EPA) are compared with the proposed interference-aware power control algorithm with convex pricing (NPG-CP). In the POCP algorithm, the transmission power of each user is based on their target SINRs. The values of their target SINRs are set to 5 dB. The simulation parameters are summarized in Table 5.1. Transmit power, users' utilities and energy efficiency of the proposed scheme are verified, compared with existing schemes.

Figure 5.2 shows that the total user utilities increase as the number of femtocell increases. The corresponding equilibrium transmission power consumption of NPG

Table 5.1 Simulation parameters of system

Parameter	Value
Macrocell radius R_m	288 m
Femtocell radius R_f	5 m
Grid size D_{grid}	100 m
Carrier frequency $f_{c,MHz}$	2 GHz
System bandwidth w	10 MHz
MBS/ FBS TX power	1/0.1 W
Thermal noise density	174 dBm/Hz
Out/In-door path loss exponent	4/3
UQS parameters	$\alpha_i = 8$, $\beta_i = 1.2$

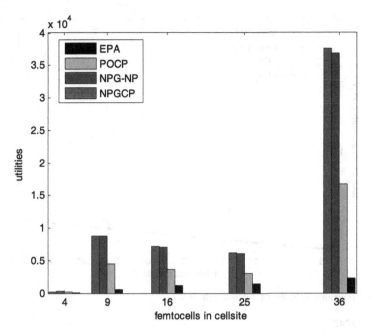

Fig. 5.2 Utilities of different schemes

schemes and power consumption of the other two schemes is displayed in Fig. 5.3. It can be seen from Fig. 5.2 that NPG-NP scheme has obvious advantage in user utility than POCP and EPA schemes. This illustrates that the user utility definition of UQS and individual transmission power is more efficient in our proposed schemes. What's more, when the number of femtocell users per cellsite is bigger, the advantage in utility of NPG-CP is more obvious than NPG-NP scheme. And the not so big difference in user utilities between NPG-CP and NPG-NP is owing to the function form of user utility. In other words, our proposed scheme can improve the sum of user utilities significantly.

Figure 5.3 represents comparison of the total transmission power of the femtocell network among different power control schemes. Obviously, the scheme NPG-NP consumes much lesser transmission power than POCP and EPA schemes. As a result of pricing, the transmit power at equilibrium in our proposed scheme NPG-CP is lower than that without pricing scheme NPG-NP. Compared with others schemes, our proposed scheme NPG-CP is most energy saving.

More detailed power consumption comparison of single user is illustrated in Fig. 5.4. The solid lines indicate the transmission power of user in NPG-CP scheme with the optimal pricing factor. And the dotted lines present the power consume of single user in NPG-NP scheme. Almost all the users consume less power in NPG-CP than NPG-NP scheme, which is consistent with the theoretical analysis. The optimal power allocation can be finished in about five iterations.

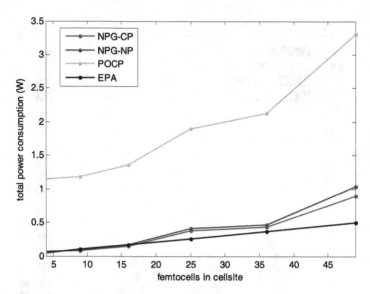

Fig. 5.3 Transmission power of different schemes

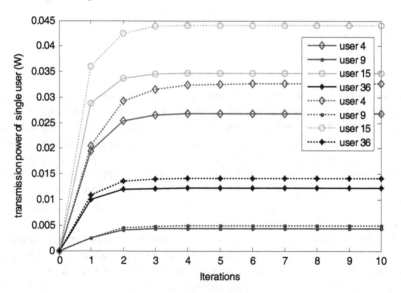

Fig. 5.4 Comparison of single user's transmission power of convex pricing scheme and unpriced scheme

5.5   Conclusion

In this chapter, we have proposed an energy-aware uplink power control scheme for two-tier femtocell networks based on non-cooperative game. We can obtain Pareto improvement of the non-cooperative power control game via cross-tier interference pricing. Based on the proposed power optimization scheme each femto user can reduce the system power, while mitigating uplink inter-cell interference. It has been shown through simulations that the proposed power allocation algorithm is able to not only maintain good user utilities but also improve energy efficiency, as compared with existing power allocation schemes.

Acknowledgements The authors would like to thank Ms. Deli Liu for her contribution and helpful discussions.

References

1. V. Chandrasekhar and J. G. Andrews, "Femtocell networks: A survey," *IEEE Commun. Mag.*, vol. 46, no. 9, pp. 59–67, 2008.
2. D. López-Pérez, A. Valcarce, G. de la Roche, and J. Zhang, "Ofdma femtocells: A roadmap on interference avoidance," *IEEE Commun. Mag.*, vol. 47, no. 9, pp. 41–48, 2009.
3. J.-H. Yun and K. G. Shin, "Adaptive interference management of ofdma femtocells for co-channel deployment," *IEEE J. Sel. Areas in Commun.*, vol. 29, no. 6, pp. 1225–1241, 2011.
4. K. Son, S. Lee,Y. Yi, and S. Chong, "Refim: A practical interference management in heterogeneous wireless access networks," *IEEE J. Sel. Areas in Commun.*, vol. 29, no. 6, pp. 1260–1272, 2011.
5. V. Chandrasekhar, J. G. Andrews, T. Muharemovic, Z. Shen, and A. Gatherer, "Power control in two-tier femtocell networks," *IEEE Trans. Wireless Commun.*, vol. 8, no. 8, pp. 4316–4328, 2009.
6. X. Kang, R. Zhang, and M. Motani, "Price-based resource allocation for spectrum-sharing femtocell networks: a stackelberg game approach," *IEEE J. Sel. Areas in Commun.*, 2012.
7. S. Barbarossa, S. Sardellitti, A. Carfagna, and P. Vec-chiarelli, "Decentralized interference management in femtocells: A game-theoretic approach," in *IEEE CROWNCOM'10*, June 2010, pp. 1–5.
8. Z. Jiang, Y. Li, "Max-utility wireless resource management for best effort traffic," *IEEE Trans. on Wireless Communications*, vol. 4, no. 1, pp. 100–111, 2005.
9. C. U. Saraydar, N. B. Mandayam, and D. J. Goodman, "Efficient power control via pricing in wireless data networks," *Communications, IEEE Transactions on*, vol. 50, no. 2, pp. 291–303, 2002.
10. D. Liu, W. Zheng, H. Zhang, et al., "Energy efficient power optimization in two-tier femtocell networks with interference pricing," in *IEEE ICCNT'12*, 2012, pp. 247–252.
11. J.-W. Lee, R. R. Mazumdar, and N. B. Shroff, "Opportunistic power scheduling for dynamic multi-server wireless systems," *Wireless Communications, IEEE Transactions on*, vol. 5, no. 6, pp. 1506–1515, 2006.
12. M. Rasti, A. R. Sharafat, and B. Seyfe, "Pareto-efficient and goal-driven power control in wireless networks: a game-theoretic approach with a novel pricing scheme," *IEEE/ACM Transactions on Networking (TON)*, vol. 17, no. 2, pp. 556–569, 2009.
13. P. Liu, P. Zhang, S. Jordan, and M. L. Honig, "Single-cell forward link power allocation using pricing in wireless networks," *Wireless Communications, IEEE Transactions on*, vol. 3, no. 2, pp. 533–543, 2004.
14. T. Alpcan, L. Pavel, and N. Stefanovic, "A control theoretic approach to noncooperative game design," in *Decision and Control, 2009 held jointly with the 2009 28th Chinese Control Conference. CDC/CCC 2009. Proceedings of the 48th IEEE Conference on*. IEEE, 2009, pp. 8575–8580.
15. R. Wilson, "Nonlinear pricing," *OUP Catalogue*, 2001.
16. D. Fudenberg and J. Tirole, *Game Theory*. MIT Press, 1993.

Chapter 6
Differentiated-Pricing Based Power Allocation in Dense Femtocell Networks

Abstract Femtocells, combined with orthogonal frequency division multiple access (OFDMA), can improve cellular coverage and offload traffics from existing macrocells. However, the co-channel deployment of femtocells is still facing challenges arising from potentially severe co-channel interference in dense femtocells. In this chapter, we investigate the uplink power allocation problem in dense deployed femtocells. We first model the uplink power allocation in femtocells as a non-cooperative game, where co-channel interference is taken into account in maximizing the femtocell capacity and uplink femto-to-femto interference is alleviated by charging each femto user a price proportional to the signal to interference and noise ratio (SINR). Based on the non-cooperative game, we then devise a distributed power allocation algorithm with differentiated pricing update. Simulation results show that the proposed power allocation algorithm is not only able to provide improved capacities for femtocells, but also can improve users' fairness, as compared with existing unpriced water filling power allocation algorithm.

6.1 Introduction

Femtocell has attracted a lot of attention for its potential in improving coverage and capacity for indoor environments. This is because most of the voice services and data traffics occur indoors, where however the coverage of macrocells is not sufficient [1]. Femtocells combined with orthogonal frequency division multiple access (OFDMA) have been considered in major wireless communication standards such as 3GPP LTE/LTE-Advanced [2]. Spectrum partition among femtocells, where femtocells are assigned with different (or orthogonal) frequency bands, may not be preferred by operators due to the scarcity of spectrum resources and difficulties in implementation. While in spectrum-sharing deployment, where femtocells share the same spectrum, co-tier interference could be severe [3], especially when femtocell base stations (FBSs) are densely deployed close to each other [4]. To mitigate the

H. Zhang et al., *4G Femtocells: Resource Allocation and Interference Management*,
SpringerBriefs in Computer Science, DOI 10.1007/978-1-4614-9080-7_6,
© The Author(s) 2013

co-channel interference between femtocells, resource allocation has been widely used in dense femtocells.

Power control has also been widely used to mitigate inter-femtocell interference in co-channel deployment of femtocells. In [4], a power control algorithm is proposed to maximize the total capacity of densely deployed femtocells while controlling the co-tier interference, but the fairness between femto users is not considered.

Recently, several studies considering pricing techniques together with power controls have been reported. In [5], the distributed cross-tier interference pricing power allocation for co-channel deployed femtocells is modeled as a non-cooperative game, but the constraint on maximum femto-user transmit power is ignored in solving the non-cooperative game. For alleviating uplink interference caused by co-channel femto users to macrocells, a distributed femtocell power control algorithm is developed based on non-cooperative game theory in [6]; while in [7] femto users are priced for causing interference to macrocells in the power allocation based on a Stackelberg model.

In this chapter, different from the previous work in [8], we focus on the uplink power allocation problem in dense deployed femtocells [9]. Moreover, each femto user is charged with a dynamically optimized price proportional to the amount of femto user's signal to interference and noise ratio (SINR). Based on the non-cooperative game, we then devise a distributed algorithm for each femtocell. Simulation comparisons with modified iterative water filling (MIWF) based power allocation show that the proposed distributed femtocell power allocation algorithm is able to provide not only improved capacities but also improved fairness in femtocell networks.

The rest of this chapter is organized as follows. The system model and problem formulation are presented in Sect. 6.2. In Sect. 6.3, the differentiated-pricing based power allocation algorithm is proposed. Performance of the proposed algorithm is evaluated by simulations in Sect. 6.4. Finally, Sect. 6.5 concludes the chapter.

6.2 System Model and Problem Formulation

6.2.1 System Model

In this chapter, a two-tier OFDMA femtocell and macrocell network is considered, in which K FBSs overlaid by a central Macro Base Station (MBS) as shown in Fig. 6.1. We focus on the power allocation in the uplink of femtocells. It is assumed that the total bandwidth B is divided into N equal-bandwidth sub-channels.

Let M and F denote the numbers of macro users and femto users, respectively. Let $f_{k,n} \in \{1, 2, \ldots, F\}$ denote the user in the kth ($k \in \{1, 2, \ldots, K\}$) femtocell using the nth ($n \in \{1, 2, \ldots, N\}$) subchannel and $m_n \in \{1, 2, \ldots, M\}$ denote the user in the central macrocell using the nth subchannel.

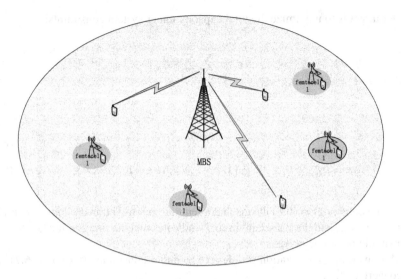

Fig. 6.1 The topology of two-tier femtocell networks

The received SINR of user $f_{k,n}$ can be expressed as:

$$\gamma_{f_{k,n}} = \frac{p_{f_{k,n}} g_{k,f_{k,n}}}{I_{f_{k,n}} + \sigma^2},$$ (6.1)

where $p_{f_{k,n}}$ is the power of femtocell user $f_{k,n}$, $g_{k,f_{k,n}}$ is the channel gain from the transmitting user $f_{k,n}$ to the receiving FBS k, σ^2 is the additive white Gaussian noise power, and $I_{f_{k,n}}$ is the interference which can be written as:

$$I_{f_{k,n}} = \sum_{j=1, j \neq k}^{K} p_{f_{j,n}} g_{k,f_{j,n}} + p_{m_n} g_{k,m_n},$$ (6.2)

where p_{m_n} is the power of macrocell user m_n and g_{k,m_n} is the channel gain from the transmitting m_n user to the central receiving MBS. Moreover, the first component on the right-hand side of the equation is the interference caused by other co-channel femtocells, and the second component is caused by the central macrocell.

The capacity that indicates the achievable instantaneous data rate of the femtocell user $f_{k,n}$ based on Shannon's formula can be expressed as:

$$C_{f_{k,n}} = \frac{B}{N} \log_2(1 + \gamma_{f_{k,n}}).$$ (6.3)

6.2.2 Problem Formulation

We define C_{tot} as the capacity of K femtocells, which can be formulated as:

$$C_{tot} = \sum_{k=1}^{K} \sum_{n=1}^{N} C_{f_{k,n}}$$ (6.4)

Our target is to maximize the total capacity under system constraints:

$$C_{tot} = \sum_{k=1}^{K} \sum_{n=1}^{N} C_{f_{k,n}} \tag{6.5}$$

s.t.

$$p_{f_{k,n}} \le p_{\max}, \quad \forall n \in \{1,2,\ldots,N\}, \forall k \in \{1,2,\ldots,K\} \tag{6.6}$$

$$p_{f_{k,n}} \ge 0, \quad \forall n \in \{1,2,\ldots,N\}, \forall k \in \{1,2,\ldots,K\} \tag{6.7}$$

where the constraint in (6.6) means that a femto user's total transmit power can't be more than $p_{\max}$, and the constraint in (6.7) indicates that the power allocated to each user should be nonnegative.

Moreover, the optimization problem (6.5) under constraints (6.6) and (6.7) can be converted to:

$$C_{tot} = \sum_{k=1}^{K} C_{f_{k,n}}, \quad \forall n \in \{1,2,\ldots,N\} \tag{6.8}$$

s.t.

$$p_{f_{k,n},n} \le p_{n,\max}, \quad \forall n \in \{1,2,\ldots,N\}, \forall k \in \{1,2,\ldots,K\} \tag{6.9}$$

$$0 \le p_{f_{k,n},n}, \quad \forall n \in \{1,2,\ldots,N\}, \forall k \in \{1,2,\ldots,K\} \tag{6.10}$$

That is because the total capacity maximization is equivalent to the maximization of the total capacity of K femtocells on each subchannel, since the power allocation is independent for each subchannel.

6.3 Differentiated-Pricing Based Power Allocation

Firstly, we propose a non-cooperative power allocation scheme based on game theory for given subchannel allocation [10]. Secondly, based on a differentiated pricing mechanism, a power allocation scheme by pricing the femtocell users according to their SINR is developed. Moreover, a dynamic power allocation algorithm is formulated.

6.3.1 Game Theoretic Based on Differentiated Pricing

Users in each coordination cell are considered as selfish and rational players. Each of them tries to maximize their individual utility without considering the corresponding interference to other co-channel users. To balance the contradiction between them, pricing mechanism is used in our proposed scheme.

Definition 1. The power allocation problem is modeled as a femtocell non-cooperative power allocation game (NPAG):

$$G = < \Gamma_n, \mathbf{P}_n, \mu_n >, \tag{6.11}$$

where $\Gamma_n = \{f_{1,n}, f_{1,n}, \ldots, f_{K,n}\}$, $\forall n \in \{1,2,\ldots.N\}$, denotes the set of the femto users (i.e., the players set) using the nth subchannel of all FBSs, $\mathbf{P}_n = \left\{ p_{f_{1,n}}, p_{f_{2,n}}, \ldots, p_{f_{K,n}} \right\}$, $\forall n \in \{1,2,\ldots.N\}$ is the power allocation strategy space of the players, and $\mu_n = \left\{ \mu_{f_{1,n}}, \mu_{f_{2,n}}, \ldots, \mu_{f_{K,n}} \right\}$, $\forall n \in \{1,2,\ldots.N\}$ is the net utility set based on the differentiated pricing mechanism.

We assume different users are charged at different prices according to the variable channel conditions, considering that the femtocell has complete information about the network. Under the differentiated pricing rule, we denote the pricing function of user $f_{k,n}$ as:

$$c_{f_{k,n}} = \frac{B\gamma_{f_{k,n}}}{N \ln 2(1 + \gamma_{f_{k,n}})}. \tag{6.12}$$

Therefore, the utility of user $f_{k,n}$ can be given as follows:

$$\mu_{f_{K,n}} = C_{f_{k,n}} - c_{f_{k,n}}, \tag{6.13}$$

where $c_{f_{k,n}} = \alpha_{f_{k,n}} \gamma_{f_{k,n}}$, and $\alpha_{f_{k,n}} > 0$ is the price coefficient charged on femto user with respect to the corresponding SINR.

Definition 2. Player $f_{k,n}$'s best response given the power allocation of all other co-channel femtocell users is expressed as:

$$\hat{p}_{f_{k,n}} = \arg\max_{p_{f_{k,n}}} \mu_{f_{k,n}} \left(p_{f_{k,n}} \middle| \mathbf{P}_{-f_{k,n}} \right), \tag{6.14}$$

where $\mathbf{p}_{-f_{k,n}} = \left\{ p_{f_{1,n}}, p_{f_{2,n}}, \ldots, p_{f_{k-1,n}}, p_{f_{k+1,n}}, \ldots, p_{f_{K,n}} \right\}$ is the power vector of co-channel femto users other than $f_{k,n}$ of subchannel n in all femtocells except for femtocell k.

Definition 3. Denote $\hat{\mathbf{P}}_n = \left\{ \hat{p}_{f_{1,n}}, \hat{p}_{f_{2,n}}, \cdots, \hat{p}_{f_{k,n}} \right\}$, if condition (6.15) is met:

$$\mu_{f_{K,n}}(\hat{p}_{f_{k,n}}, \hat{\mathbf{P}}_{-f_{k,n}}) \geq \mu_{f_{K,n}}(p_{f_{k,n}}, \hat{\mathbf{P}}_{-f_{k,n}}), \ \forall p_{f_{k,n}} \in \mathbf{P}_n, \tag{6.15}$$

then $\hat{\mathbf{P}}_n = \left\{ \hat{p}_{f_{1,n}}, \hat{p}_{f_{2,n}}, \cdots, \hat{p}_{f_{k,n}} \right\}$ is the optimal Nash Equilibrium (NE) transmitting power vector of the co-channel users. NE is defined as the fixed point where no player can improve their utility by changing its strategy unilaterally [10].

In the following, we will prove the existence and uniqueness of NE in the NPAG.

Theorem 1. An NE exists in the NPAG.

Proof. A NE exists in the NPAG if the following two conditions are satisfied according to the Nash theorem:

1. In the finite Euclidean space R^K, $\mathbf{P}_n$ is non-empty, convex and compact.
2. μ_n is continuous and concave with $\mathbf{P}_n$.

Since the allocated power on each subchannel should be more than zero and less than the maximum p_{max}, condition 1 is obviously satisfied.

Next we will prove the condition 2. Equation (6.13) can be rewritten as:

$$\mu_{f_{K,n}} = \frac{B}{N} \left[\log_2\left(1 + \frac{p_{f_{k,n}} g_{k,f_{k,n}}}{\sum\limits_{j=1,j\neq k}^{K} p_{f_{j,n}} g_{k,f_{j,n}} + p_{mn} g_{k,mn} + \sigma^2}\right) - \frac{\alpha_{f_{k,n}} p_{f_{k,n}} g_{k,f_{k,n}}}{\ln 2 (\sum\limits_{j=1,j\neq k}^{K} p_{f_{j,n}} g_{k,f_{j,n}} + p_{mn} g_{k,mn} + \sigma^2)} \right] \tag{6.16}$$

It is obvious that $\mu_{f_{K,n}}$ is continuous with $\mathbf{P}_n$. And the first order derivative of the femto user $f_{k,n}$'s utility function with respect to $p_{f_{k,n}}$ is as follows:

$$\frac{\partial \mu_{f_{k,n}}}{\partial p_{f_{k,n}}} = \frac{B}{N \ln 2} \left[\frac{g_{k,f_{k,n}}}{\sum\limits_{j=1,j\neq k}^{K} p_{f_{j,n}} g_{k,f_{j,n}} + p_{mn} g_{k,mn} + \sigma^2 + p_{f_{k,n}} g_{k,f_{k,n}}} - \frac{\alpha_{f_{k,n}} g_{k,f_{k,n}}}{\sum\limits_{j=1,j\neq k}^{K} p_{f_{j,n}} g_{k,f_{j,n}} + p_{mn} g_{k,mn} + \sigma^2} \right] \tag{6.17}$$

If $\frac{\partial \mu_{f_{k,n}}}{\partial p_{f_{k,n}}} \geq 0$, then we can get:

$$\alpha_{f_{k,n}} \leq \frac{B}{N \ln 2 \left(1 + \dfrac{p_{f_{k,n}} g_{k,f_{k,n}}}{\sum\limits_{j=1,j\neq k}^{K} p_{f_{j,n}} g_{k,f_{j,n}} + p_{mn} g_{k,mn} + \sigma^2}\right)} \tag{6.18}$$

Note that, $\alpha_{f_{k,n}} \in {}^{+}$, therefore:

$$0 \le \alpha_{f_{k,n}} \le \cfrac{B}{N \ln 2 \left(1 + \cfrac{P_{f_{k,n}} g_{k,f_{k,n}}}{\sum\limits_{j=1,j\neq k}^{K} P_{f_{j,n}} g_{k,f_{j,n}} + P_{mn} g_{k,mn} + \sigma^2} \right)} \tag{6.19}$$

The second order derivative can be formulated as:

$$\frac{\partial^2 \mu_{f_{k,n}}}{\partial^2 P_{f_{k,n}}} = -\cfrac{B g_{k,f_{k,n}}^2}{N \ln 2 \left(\sum\limits_{j=1,j\neq k}^{K} P_{f_{j,n}} g_{k,f_{j,n}} + P_{mn} g_{k,mn} + \sigma^2 + P_{f_{k,n}} g_{k,f_{k,n}} \right)^2} \tag{6.20}$$
$$\le 0$$

It is seen that the second order derivative is nonpositive. Therefore, $\mu_{f_{K,n}}$ is a quasi-concave function of $p_{f_{k,n}}$.

Both 1 and 2 are satisfied, we can know that a NE exists in the NPAG.

Theorem 2. The NPAG has a unique NE.

The detailed proof can be found in [10].

Based on the above analysis, we can say a NE exists in the NPAG. That is to say, $\widehat{\mathbf{P}}_n$ is the optimal power allocation solution.

6.3.2 Differentiated-Pricing Function

Differentiated pricing is also referred to as price discrimination in the economics literature [11]. As mentioned above, different users are charged at different prices. Moreover, it is depending on SINR which can indicate the variable channel conditions. In the following analysis we will give the detail expression of $\alpha_{f_{k,n}}$.

Proof. Assume that $\alpha_n = \{ \alpha_{f_{1,n}}, \alpha_{f_{2,n}}, \ldots, \alpha_{f_{K,n}} \}$ is the differentiated pricing vector, which can be expressed in terms of the corresponding allocated power and the variable channel conditions as follows:

$$\alpha_{f_{k,n}} = \cfrac{B}{N \ln 2 \left(1 + \cfrac{P_{f_{k,n}} g_{k,f_{k,n}}}{\sum\limits_{j=1,j\neq k}^{K} P_{f_{j,n}} g_{k,f_{j,n}} + P_{mn} g_{k,mn} + \sigma^2} \right)} \tag{6.21}$$

Proof. From (6.19), we know the value range of $\alpha_{f_{k,n}}$. It is obvious that the value of $\alpha_{f_{k,n}}$ in (6.21) is the maximum value in (6.19). Therefore, Proposition 6.3.2 is proved.

Lemma 1. The best power allocation response of the NPAG in (6.14) can be expressed as:

$$\hat{P}_{f_{k,n}} = \left[\frac{B(\sum_{j=1,j\neq k}^{K} P_{f_{j,n}} g_{k,f_{j,n}} + P_{mn} g_{k,mn} + \sigma^2)}{N \ln 2 \alpha_{f_{k,n}} g_{k,f_{k,n}}} - \frac{\sum_{j=1,j\neq k}^{K} P_{f_{j,n}} g_{k,f_{j,n}} + P_{mn} g_{k,mn} + \sigma^2}{g_{k,f_{k,n}}} \right]_0^{P_{max}}$$
(6.22)

Proof. Given the corresponding $\alpha_{f_{k,n}}$ according to (6.21), by seeing (6.17) equal to 0, we can get the optimal power allocated to femto user $f_{k,n}$ in femtocell k at subchannel n as expressed in (6.22). In the formulation, $[c]_a^b$ means minmaxa, c, b.

6.3.3   Dynamic Power Allocation

The uniqueness of NE is proved in the above analysis. In this subsection, an optimal dynamic iterative power allocation algorithm is proposed to converge to NE, as shown in algorithm 1.

Algorithm 8 Iterative algorithm for dynamic power allocation

1: Initialize subchannel set: $\mathbf{N} = \{1, 2, \ldots, N\}$, Femto User set: $\mathbf{F} = \{1, 2, \ldots, F\}$, FBS set: $\mathbf{K} = \{1, 2, \ldots, K\}$
2: Collect the channel gain:
   $g_{k,f_{k,n}} (\forall n \in \mathbf{N}, \forall k \in \mathbf{K})$, the channel gain from the transmitting user $f_{k,n}$ to the receiving FBS k;
   $g_{k,m_n} (\forall n \in \mathbf{N}, \forall k \in \mathbf{K})$, the channel gain from the transmitting m_n user to the central receiving MBS.
3: Calculate SINR based on the channel gain for each user;
4: Each femtocell compute its optimal power based on the received price $\alpha_{f_{k,n}}$ by (6.22);
5: Each femtocell update its price according to (6.21);
6: Step (6.4) and Step (6.5) are repeated until convergence.

6.4   Simulation Results and Discussion

6.4.1   Simulation Parameters

In the subsection, we present simulation results to evaluate the performance of the proposed NPAG algorithm, as compared with MIWF power control algorithm [12]. Both the system capacity and the fairness are evaluated in the simulations.

In the simulations, the macrocell has a coverage radius of 500 m. Each femtocell has a coverage radius of 10 m. K FBSs and 50 macro users are randomly distributed in the macrocell coverage area. The minimum distance between the MBS and a macro user (or an FBS) is 50 m. The minimum distance between FBSs is 40 m. Femto users are uniformly distributed in the coverage area of their serving femtocell.

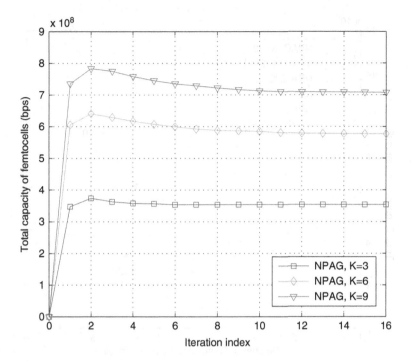

Fig. 6.2 Convergence versus the number of iterations

Both macro and femto cells employ a carrier frequency of 2 GHz, B = 10 MHz, and N = 50. The AWGN variance is given by $\sigma^2 = \frac{B}{N}N_0$, where $N_0 = -174$ dBm/Hz. The Rayleigh-fading channel gains are modeled as unit-mean exponentially distributed random variables. The average channel gain (including pathloss and antenna gains) for indoor femto user and outdoor macro user are modeled as λd^{-4} and λd^{-3}, respectively, where $\lambda = 2 \times 10^{-4}$[6]. The maximum uplink transmission powers of a femto user and a macro user is set as 20 and 30 dBm, respectively.

6.4.2   Performance Analysis

Figure 6.2 shows the convergence of the proposed algorithm in terms of the average capacity per femtocell versus iterations. We can see that the proposed algorithm takes only 10 iterations to converge, indicating that it is suitable for real-time implementation.

Figure 6.3 shows the total capacity of the femtocell when the number of femto users per femtocell increases from 1 to 6, for K = 20, 30, and 50. It can be observed that the proposed NPAG algorithm outperforms the MIWF algorithm by up to a 10 %

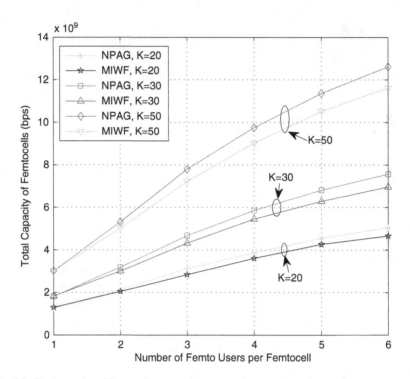

Fig. 6.3 Total capacity of femtocells

increase in total femtocell capacity. As the number of femtocells K increases, the advantage of the NPAG algorithm becomes more noticeable, because the increased co-tier uplink interference can be effectively mitigated by the differentiated pricing scheme imposed on femto users in the NPAG algorithm, but not by the MIWF based algorithm.

Figure 6.4 compares the average fairness between femto users in each femtocell between the proposed NPAG algorithm and the MIWF algorithm. The fairness metric is based on the classic Jain's fairness index (JFI) [13]. It can be seen from the figure, as the number of the users in each femtocell increases, the fairness in FPAG is better than MIWF method.

6.5   Conclusion

In this chapter, we proposed a differentiated-pricing based power allocation algorithm for the uplink frequency-sharing femtocells, based on the non-cooperative game framework. Using the proposed algorithm, each femtocell can maximize its

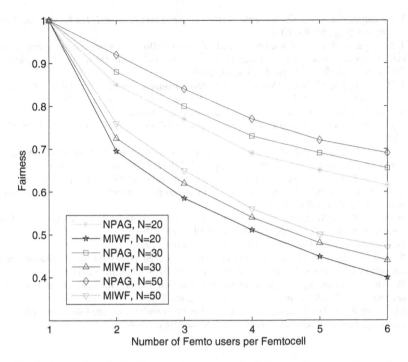

Fig. 6.4 Total capacity of all delay sensitive users of all femtocells versus the number of femtocells K

capacity through power allocation, and with uplink femto-to-femto interference alleviated by differentiated pricing functions imposed on femto users. It has been shown through simulations that the proposed power allocation algorithm is able to provide improved capacities of femtocells, together with better fairness, as compared with the existing unpriced MIWF based power allocation algorithm.

Acknowledgements The authors would like to thank Ms. Wenmin Ma and Dr. Wei Zheng for their contributions and helpful discussions. This research has been supported by National Key Technology R&D Program of China (2010ZX03003-001-01, 2011ZX03003-002-01), the National Natural Science Foundation of China (61101109), the Co-building Project of Beijing Municipal Education Commission "G-RAN based Experimental Platform for Future Mobile Communications" and "Research on Resource Allocation and Scheduling Strategy of Future Wireless Communication System" and Key Fund of Beijing Key Laboratory on Future Network Research.

References

1. D. López-Pérez, A. Valcarce, G. de la Roche, and J. Zhang, "Ofdma femtocells: A roadmap on interference avoidance," *IEEE Commun. Mag.*, vol. 47, no. 9, pp. 41–48, 2009.
2. *E-UTRA and E-UTRAN Overall Description*, 3GPP Std. TS 36.300 v10.0.0, 2010.

3. V. Chandrasekhar and J. G. Andrews, "Femtocell networks: A survey," *IEEE Commun. Mag.*, vol. 46, no. 9, pp. 59–67, 2008.
4. J. Kim and D.-H. Cho, "A joint power and subchannel allocation scheme maximizing system capacity in indoor dense mobile communication systems," *IEEE Transactions on Vehicular Technology*, vol. 59, no. 9, pp. 4340–4353, 2010.
5. J.-H. Yun and K. G. Shin, "Adaptive interference management of ofdma femtocells for co-channel deployment," *IEEE J. Sel. Areas in Commun.*, vol. 29, no. 6, pp. 1225–1241, 2011.
6. V. Chandrasekhar, J. G. Andrews, T. Muharemovic, Z. Shen, and A. Gatherer, "Power control in two-tier femtocell networks," *IEEE Trans. Wireless Commun.*, vol. 8, no. 8, pp. 4316–4328, 2009.
7. X. Kang, R. Zhang, and M. Motani, "Price-based resource allocation for spectrum-sharing femtocell networks: a stackelberg game approach," *IEEE J. Sel. Areas in Commun.*, 2012.
8. H. Zhang X. Chu, W. Zheng and X. Wen, "Optimization method for the joint allocation of modulation schemes, coding rates, resource blocks and power in self-organizing lte networks," in *ICC, 2012 Proceedings IEEE*, 2012.
9. W. Ma, H. Zhang, W. Zheng, X. Wen, "Differentiated-Pricing Based Power Allocation in Dense Femtocell Networks", to appear in Proc. of IEEE WPMC 2012.
10. D. Fudenberg and J. Tirole, *Game Theory.*   MIT Press, 1993.
11. S. Ren and M. van der Schaar, "Pricing and distributed power control in wireless relay networks," *IEEE Trans. Sig. Proc.*, vol. 59, no. 6, pp. 2913–2926, June 2011.
12. W. Yu, "Sum-capacity computation for the gaussian vector broadcast channel via dual decomposition," *IEEE Trans. Inf. Theory*, vol. 52, no. 2, pp. 754–759, Feb. 2006.
13. M. C. Erturk, H. Aki, I. Guvenc, , and H. Arslan, "Fair and qos-oriented spectrum splitting in macrocell-femtocell networks," in *IEEE Globecom'10*, pp. 1–6.

Chapter 7
Conclusions and Future Works

Abstract In this book, we investigate resource allocation and interference management in femtocell networks. A subchannel allocation algorithm based on the ant colony optimization is proposed in dense deployed femtocells. To explore the multi-dimensional diversity in resource allocation, we propose a semi-distributed interference-aware resource allocation algorithm for the uplink of co-channel deployed femtocells, based on a non-cooperative game framework. Furthermore, a joint subchannel and power allocation scheme for spectrum-sharing femtocells is investigated taking heterogeneous femto-user QoS requirements and a cross-tier interference limit into account. In Chap. 5, we have proposed an energy-aware uplink power control scheme for two-tier femtocell networks based on non-cooperative game. In Chap. 6, we have proposed a differentiated-pricing based power allocation algorithm for the uplink frequency-sharing femtocells, based on the non-cooperative game framework.

7.1 Conclusions

In this book, we investigate resource management and interference mitigation in femtocell networks. In Chap. 1, we give a introduction for 4G femtocells, where we give a survey for the research of resource allocation and interference management. Interference mitigation based on power control and subchannel scheduling is surveyed. Moreover, resource allocation based on game theory and convex optimization is also surveyed in Chap. 1, where challenges in resource allocation and interference management are given.

In Chap. 2, subchannel allocation algorithms based on Ant Colony Algorithm are presented. As a typical meta-heuristic method, ACA provides simple and robust way for resource allocation optimization. We formulate the resource allocation problem as path searching in a graph, and use the pheromone trail and the heuristic

H. Zhang et al., *4G Femtocells: Resource Allocation and Interference Management*,
SpringerBriefs in Computer Science, DOI 10.1007/978-1-4614-9080-7__7,
© The Author(s) 2013

information to guide the construction of solution construction. In comparison with the traditional RR algorithm, better throughput can be achieved by ACA based algorithms. ACA-MF and ACA-FF can guarantee the fairness among users while meeting rate requirements.

In Chap. 3, we have proposed a semi-distributed interference-aware resource allocation algorithm for the uplink of co-channel deployed femtocells, based on a non-cooperative game framework. Using the proposed algorithm, each femtocell can maximize its capacity through resource allocation, taking into account inter-cell interference reported by its femto users, and with uplink femto-to-macro interference alleviated by a pricing scheme imposed on femto users. It has been shown through simulations that the proposed interference-aware resource allocation algorithm is able to provide improved capacities of both macrocell and femtocells, together with comparable tiered fairness, as compared with the existing unpriced subchannel allocation and MIWF based power allocation algorithm.

In Chap. 4, we have investigated the joint subchannel and power allocation for spectrum-sharing femtocells taking heterogeneous femto-user QoS requirements and a cross-tier interference limit into account. The proposed distributed resource allocation algorithm properly allocates resources to users according to their heterogeneous QoS requirements, so as to increase the throughout in the network. Simulation results have shown that the proposed algorithm provides more efficient solutions compared with cutting-edge algorithms in the literature.

In Chap. 5, we have proposed an energy-aware uplink power control scheme for two-tier femtocell networks based on non-cooperative game. We can obtain Pareto improvement of the non-cooperative power control game via cross-tier interference pricing. Based on the proposed power optimization scheme each femto user can reduce the system power, while mitigating uplink inter-cell interference. It has been shown through simulations that the proposed power allocation algorithm is able to not only maintain good user utilities but also improve energy efficiency, as compared with existing power allocation schemes.

In Chap. 6, we proposed a differentiated-pricing based power allocation algorithm for the uplink frequency-sharing femtocells, based on the non-cooperative game framework. Using the proposed algorithm, each femtocell can maximize its capacity through power allocation, and with uplink femto-to-femto interference alleviated by differentiated pricing functions imposed on femto users. It has been shown through simulations that the proposed power allocation algorithm is able to provide improved capacities of femtocells, together with better fairness, as compared with the existing unpriced MIWF based power allocation algorithm.

7.2 Future Works

Though we have done some works in resource allocation and interference management in femtocell networks, there are still some challenges and open questions.

For the future works, we will investigate the other techniques for interference management in macro-femto networks, such as beamforming, clustering method to mitigate the co-tier and cross-tier interference in heterogenous femtocell networks. Moreover, the cross-tier interference temperature can be extended to the co-tier interference mitigation. For the resource allocation aspects, cross-layer resource scheduling involving the Modulation and Coding Scheme (MCS) at the PHY layer, together with the power and subchannel allocation will be jointly optimized in femtocell networks. Moreover, resource allocation and interference mitigation based on inter-femtocell coordination will be promising. Finally, the resource allocation and interference management proposed in the book can be easily extended to the other small cell networks, such as picocells or cognitive relay networks.